THE PHYSICS OF RUC

Nottingham University Press
Manor Farm, Main Street, Thrumpton
Nottingham, NG11 0AX, United Kingdom
www.nup.com

NOTTINGHAM

First published 2009

British Library Cataloguing in Publication Data

The Physics of Rugby
TD Lipscombe

ISBN 978-1-904761-17-4

Cover photo by Jusben (www.morguefile.com)

Typeset by Nottingham University Press, Nottingham
Printed and bound by Hobbs the Printers, Hampshire, England

The Physics of Rugby

Trevor Davis Lipscombe

Table of Contents

INTRODUCTION

'Rugby is a good occasion for keeping thirty bullies far from the centre of the city.'

Oscar Wilde

'I prefer rugby to soccer. I enjoy the violence in rugby, except when they start biting each other's ears off.'

Elizabeth Taylor

This book began three decades ago, on a cold Saturday morning in early September. That was when I pulled on for the first time the scarlet jersey of Icknield, a high school named for a fierce tribe of Britons, the Iceni, who under their warrior queen Boadicea inflicted several heavy defeats upon the Romans. Our opponents that morning were from the Cardinal Newman School, named not after a warrior but a mild-mannered man who may soon become a saint of the Roman Catholic Church. The Cardinal's men were victorious that day, the final score was 10-6 and, to show my age, we were defeated by two converted tries to two unconverted tries. But I was hooked.

Chances are you are reading this book because you, too, fell in love with rugby. It has that effect on people. It's a game of many parts. While a good deal of brawn is needed, brains play a big role too, and the post-game social life is usually enjoyable, if not entirely wholesome or suitable for family viewing.

Rugby has grown immensely in recent years. There are now women's teams at the international level and a women's rugby world cup; there are national teams for the hearing impaired, and the sport spreads ever wider. As I write, Brazil conquered Trinidad and Tobago in the early qualifying stages of the next Rugby World Cup. There's even a movement to reintroduce rugby in the Olympic Games, where the current reigning champions are the United States who defeated the French 17-3, in Paris, in 1924.

For those whose playing days are over, the big screen gives us *Murderball*, which documents the 2002 World Wheelchair Rugby Championship, won by Canada. (The current world champions are the United States, who beat the Kiwis 34-30 in Christchurch). The small screen, your PC, allows you to play virtual rugby through games such as EA Sports's Rugby08, as well as head up your own team through one of the many online fantasy rugby options. And the medium screen, TV, now features high-definition picture and surround-sound audio which, combined with referees wired for sound, make watching rugby more pleasurable than ever before.

These are exciting times to be a spectator, whether in the flesh or via television. And if the gogglebox is not your thing, there are countless books on rugby to be read — some of which are included in the suggested reading section. This book, though, is slightly different.

The second passion of my life is as far removed from rugby as you can hope to get. Rugby sparks almost endless conversations about great games you've seen, or similar playing experiences that you can share. Physics, on the other hand, is a conversation stopper. To mention that you study physics generates the response 'I was never any good at that in school' and the person will look into the far distance, silently, before drifting away to find someone else to talk to.

The difficulty with physics is that, for so many people, it isn't. I mean that large numbers of textbooks and teachers give students all the details of a problem and ask them to solve it. That's not physics; it's mathematics. In a physics laboratory practical, you aren't usually asked to come up with an experiment to measure something; the equipment is all there and all you have to do is to take the data. To paraphrase Lord Rutherford of Nelson, that's not physics; it's stamp collecting.

In school or college, there's an odds on chance that you will have to analyse a rough block on an inclined plane. Who cares? But wait one moment. When two packs struggle for supremacy in the scrum, what happens if the pitch is not level? The weight force of both packs is aimed right at the centre of planet Earth. This means that some small part of the weight force acts directly downhill. Physics tells you, surprise surprise, that it's going to be easier to push downhill than uphill. More important, it shows how you can model something as complex as two groups of humans shunting and grunting as a simple block of wood on an inclined plane. Suddenly the science has a real application and the student has done something that lies at the very core of physics: take a subject from the real world and approximate it by a simple, easily solvable system. The pupil has constructed a physical model to be tested and that, in my opinion, is what physics is all about.

In the pages that follow, I try to paint a picture of rugby using the colours of physics. Throughout, I hope to show how rugby can provide great practice in creating real-world physical models, and how to use dimensional analysis and order of magnitude estimates to check them. Conversely, physics can also show you how to play better rugby. Newton, and in a sense Einstein, can help you improve your game; I've added pointers to improve your rugby wherever I can.

The language of physics is mathematics. As you journey through the book from scrums, to backs who run, to backs who tackle, to those who kick, and to game day itself, there will be a few equations along the path. Only the simplest equations are contained in the main text. If you are studying for GCSE or A-level physics – or did so many years ago – you should be able to follow the text readily. There is a second layer. Gathered in the book are notes that go to a higher mathematical level. If you delight in calculus, you will find a treatment of the physics that suits your needs in those notes sections. At the end of the book, I've collected various suggestions for further reading and I've included a glossary of physics terms in case help is needed. There's also an index of player names, so you can see if your favourite player has been included.

My love for physics took me from an undergraduate at Queen Mary College, part of the University of London, to Magdalen College at Oxford. One Wednesday in November, I was riding my bike down the Iffley Road. I had just passed the track where Sir Roger Bannister ran the first four-minute mile when I had a premonition that I'd break something during the game that afternoon. As I rode along, I wasn't particularly bothered. I'd played rugby for about 11 years and had only one black eye and a pulled hamstring to show for it. An arm or a leg wouldn't be too much of a problem, I remember thinking. Within the hour I was in the back of an ambulance being moved,

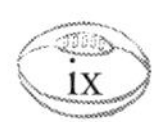

slowly and carefully, to the John Radcliffe hospital, for my neck was broken. The jersey I wore had to be cut off me (I still have it) and for close to nine months I was in clamped into various devices that resembled the medieval torture machines used in the Tower of London.

Those nine months, it turns out, were among the most productive of my life – at least in terms of physics. With nothing to do but think, I could concentrate on my research topic, the physics of flowing fluids. When I was finally able to return to Oxford, I completed my doctoral thesis swiftly, thanks to a career-ending rugby injury from which, thankfully, I completely recovered.

In thinking back on those events recently, I wondered whether there might be another, less painful, way to blend physics and rugby together. The seeds for this book were sown. Certainly physics has been applied to many different sports – baseball, basketball, ice hockey, golf, and soccer to name but a few. Rugby, to my mind, is a better sport to explain using physics than many others. For, in addition to simple mechanics, there are certain features from thermodynamics, statistics, fluid mechanics, and acoustics as well. I have tried to broaden the scope of the physics, to include as many branches of the subject as I reasonably can. For those who love physics, I trust the different topics and analogies will add to your enjoyment.

I've focused exclusively on Rugby Union, I should point out. Rugby League, while I loved listened to Eddie Waring's commentaries, is not quite the same for a physicist. There are no lineouts, rucks or mauls, and the scrums are almost a different process. I hope fans of the thirteen-a-side game will forgive me for not including the names and games of some of their truly great players.

For those who prefer Rugby Union, some explanation is in order. My brother hails from Wales, spending his early years in Grangetown, a section of Cardiff not too far from the Millennium Stadium and where you're never too far from the River Taff. I'm a Londoner. When it comes to rugby, our opinions differ. After most Six Nations games, John and I will be on the phone, chatting about the games and results. I always want England *and* Wales to win but, when they play each other, I hope the best team wins. My brother, in sharp contrast, has two favourite teams: Wales, and anyone

playing England. So, in this book, examples from Rugby Union are drawn from all of the Six Nations but are slightly biased towards the Welsh and English. *Mea culpa.*

Finally, it is with pleasure and gratitude that I thank all the people who have made this book a reality. First in the line of fire was my family. My wife Kathy and our five wonderful children Mary, Ann, Clare, Therese, and Peter have heard me talk about this book almost continually for an aeon and never once told me to shut up. I've been able, purely for research purposes of course, to huddle over a warm computer and listen to Six Nations games, and they have let me do so without complaint. I promise to be more attentive in future – at least until the start of next season.

My big brother, who had his own career-ending rugby injury, has been an invaluable rugby consultant. Not only have we analysed Six Nations campaigns and reminisced fondly about great players of the past, we've had a wonderful time during my infrequent jaunts back to Britain going to see Maesteg play in a torrential downpour and the Ospreys play at Twickenham (thanks for the tickets!) in a heat wave. My sister-in-law Jackie and my nieces Abi and Bethan made those games special.

A debt of gratitude must be paid to Mark Denny, who swiftly and skillfully went though the manuscript offering helpful suggestions, corrections, and the occasional 'you've got to be kidding.' Mark is a talented writer and an accomplished physicist, and the foremost authority on the physics of curling. His smoothing of my prose is greatly appreciated. I also deeply appreciate his analysis of 'the typical audience for this book' which, though accurate and laugh-out-loud funny, is unrepeatable in polite society.

This book has stolen into print thanks to the publication team of the Nottingham University Press (NUP), whose local rugby team produced such internationals as Rob Andrew, Dusty Hare, and Chris Oti. I am grateful to the staff at NUP, particularly Cliff, Ros, and Sarah for seeing something of value in the project.

Rudolph Zanella was patient and skillful in composing the figures that appear in the book. He took my vague suggestions and made them a reality, and never complained when I asked for yet another figure to be redrawn. I am deeply grateful for his help.

And finally, I wish to thank an anonymous rugby referee. Over the years I've often complained about, and to, the men with whistles. But to the Oxford Society referee who refused to move me when injured and insisted that an ambulance be called, I will never be able to express my thanks adequately. It is literally true that without you, this book could never have been written.

Chapter One

Pack Animals: On scrums, lineouts, rucks and mauls

'No leadership, no ideas. Not even enough imagination to thump someone in the line-up when the ref wasn't looking.'

J.P.R. Williams

'Forwards are the gnarled and scarred creatures who have a propensity for running into and bleeding all over each other.'

Peter Fitzsimmons

In Wales, rugby is a religion. As historian Trevor Fishlock wrote, Cardiff Arms Park was the altar on which 'Englishmen are sacrificed every Saturday afternoon.' The Millennium Stadium, on the banks of the Taff, is their new 56,000 tonne cathedral. Back in the seventies, the men in red were unstoppable, winning or sharing seven of the Five Nations championships that decade. The tables were turned when furious, free-flowing rugby was introduced by France. Blond bombshell flanker Jean-Pierre Rives (whose nicknames included 'Asterix' and 'the Golden Helmet' because of his shock of hair) and players such as centre Philippe Sella — the most capped player in the history of rugby, with 111 pieces of headgear to his credit — rocked the world of European rugby. No matter where you were on the pitch, Les Bleus were there in force to stop you. The French style became the model for modern rugby — almost.

England responded by playing rugby that was neither pretty, nor entertaining — but it was effective. In 1980, they won the Five Nations competition, as it then was, completing the Grand Slam. It was a style of play — known as 'grunt and shove' or 'thugby' — that still permeates English rugby, but which won them the Rugby World Cup in 2003 and earned the Runners Up spot in 2007.

The English pack in 1980 was immense. The hooker, Peter Wheeler, weighed in at 14 stone or, for the metric, close to 90kg. Wheeler was ably assisted in the front row by Phil Blakeway and Fran Cotton, neither of whom were lightweights and who, in the tradition of great props, showed no visible signs of a neck. In the second row, captain Bill Beaumont joined forces with Maurice Colclough, who played club rugby for the French club Angoulême. At number eight was Roger Uttley, one of the few people of that era to be a worthy opponent for Merve 'the Swerve' Davies, his Welsh counterpart.

All in all, England's front eight was formidable. But once they scrummed down, it wasn't so much weight, but skill and training that let them push the other four nations off of the ball. From a physics perspective, it wasn't the laws of rugby that let 'Mighty England' dominate; it was the laws of motion.

SCRUM SCIENCE

Back in 1687, Isaac Newton published his book now known as *The Principia.* It is one of the most famous scientific works ever published and it revolutionised the way scientists thought. For the first time, mathematics could fully enter the realm of physics. Newton had broad interests. He was keen on theology, on which his views were non-conformist, and he wrote much on alchemy, especially in his later years. For a time, he lived in the Tower of London while serving as master of the Royal Mint, one of the reasons why his image adorned the much-lamented British one-pound note.

In mathematics, Newton invented a branch of mathematics known as calculus. Independently, in Europe, Gottfried Leibniz did the same thing, and a bitter priority dispute broke out between the two men. The Royal Society of London — which remains the world's most exclusive club for scientists and of which Newton was a founder member — instituted a committee to investigate. They decided, to no-one's surprise, that Newton was first. Today, though, the calculus taught in high schools more closely resembles the Liebniz version.

Calculus, for a physicist, is a tool. It allows you to describe how things change, whether over time or over distance. With a way to express rates of change – how the position of Ireland's Ronan O'Gara, Wales' Shane Williams, or France's Thomas Castaignède changes over the course of the full eighty minutes, for example -- it is possible to explore the laws of motion and their consequence.

Back in the fifteenth century, Italian scientist Galileo Galilei rightly observed that two balls of different mass fall at the same rate, but he could not predict how long it would take them to fall. Newton, thanks to his equations, could. All told, Isaac came up with three laws of motion that physicists, engineers, and rugby players still use every day.

Of drop goals and pushover tries: Newton's first law

Newton's first law is that a body remains at rest, or travels in a straight line at constant speed, unless acted upon by a force. Speaking off the record, this makes a force either a push or a pull. There is some disagreement between physicists and philosophers about whether this is merely a definition of a force, or whether it actually has information content. When you've just scrummed down against the legendary Pontypool front row —Messrs Faulkner, Windsor, and Price, immortalised in song by Max Boyce — neither physicists nor philosophers are foremost in your mind. Newton's first law, however, can help you out.

Suppose you've been playing in a dour battle of forwards. Perhaps the weather is so cold that the backs can't catch the ball because their poor, delicate hands are frozen. It's been ruck, maul, and scrum the entire afternoon. Your team pushes the opposition off the ball every time, but the backs haven't capitalised on your dominance up front. Time is running out, there's a scrum on your opponent's 22, and you have the put in. What to do?

Newton says that the body (the entire scrum) stays put unless it is acted upon by a force. Every prop, second row, number eight, and flanker on the field makes some effort and exerts some sort of force. If not, they should not be on the team. If each pack exerts an equal yet opposite force, the scrum goes nowhere. That's exactly what you want, for your number nine gets to work with a lovely scrum, square and static, and can feed the ball to the outside half whenever he feels pleased to do so. The pack can break up just in time to watch the ball sail over the bar for the game-winning drop kick. Such a stable scrum helped Italy's Diego Dominguez kick three drop goals against Scotland back in 2000, equaling the record for the most scored in a Six Nations game. South Africa's Jannie de Beer smacked five of them against England in the 1999 Rugby World Cup. Mind you, for sheer power, Japan's Andrew Miller takes some beating. When playing against Fiji in the 2003 Rugby World Cup, he thumped one over the bar from an impressive 52 metres out.

Sometimes, you don't even need to score the drop goal to win; the attempt has merit, no matter what the outcome. With time running out in the 2005 Six Nations game, Wales led England by the slimmest of margins. The Welsh won a scrum close to the English line and went for the drop goal. Even though the pack was square and stationary, the kick went wide. More important, though, the ball sailed on towards the River Taff. Getting a replacement ball ate up valuable seconds on the clock, helping secure the first of a memorable series of victories for the Welsh, who ended up with their first Grand Slam in over a quarter of a century. Three years later, they'd do it again.

In real life there's a wrinkle to Newton's first law. Your push in the scrum is generated by muscles. Muscles can exert force, and the more time you spend lifting weights, the bigger this force can be. The force a muscle exerts, however, depends on the speed at which it lengthens or contracts. If your team is being pushed back a bit, your body may get compressed a tad and your leg muscles contract slightly. But when your muscles contract, they usually exert a force that is roughly constant. As your muscles lengthen, though, the force they exert trails off. So, when you scrum down, you might exert more push than your opponents – at least to begin with. Their muscles contract, offering a constant resistance. Your muscles extend, so that the force your pack exerts starts to drop off. Eventually, the two forces are equal and opposite. Newton tells us that there's no net force, so the scrum should go back at constant speed. This leads to another rugby-related manifestation of Newton's first law — the pushover try.

Suppose there's a five-metre scrum and, as before, your pack is dominant. Snap shove as usual, to move the scrum forward. Then ease off, so your push matches that of the opposition. Newton tells us the scrum will move forward at a constant speed, because no net force acts. Make sure that you merely *balance* the opponent's force; make no effort to exceed it. The scrum speed will then be constant, so your second row or number eight can easily control the ball with a large-sized boot. Someone from the powerhouse of the scrum can fall on the ball as it crosses the line; or step over it and let the glitz-and-glamour boys have the glory; or pick the ball up early and drive for the line himself. As Bill Beaumont, the former England captain and second row said of his only international try, 'I was always lethal from one yard.'

PUSH THEM OFF THE BALL: NEWTON'S SECOND LAW

The juice of Newtonian physics is mostly contained in Isaac's second law. There he speaks about momentum, which is what you get when multiplying mass by velocity. To use trade jargon, momentum and velocity are vectors, while mass is a scalar. Scalars and vectors are different animals in the zoo of physics. Scalars, such as mass, temperature, and the like, have size and size alone. A vector, on the other hand, possesses a size *and* a direction. If France's (and Stade Français) winger Christophe Dominici runs at 10 metres per second, that's a scalar quantity -- his speed. At this speed, Christophe sprints the entire length of the pitch in 10 seconds, which is quite believable –it's close to the world record of Olympic athletes. Dominici is a pink and blue blur running towards the tryline. By contrast, if he legs it at 10 m/s towards the left-hand corner flag, we're dealing with a vector (size + direction), the velocity.

Newton's second law says that force is the rate of change of momentum. This means that force, too, is a vector. Sometimes, though, no net force acts and in that case, the momentum is unchanged. This sets the rules of engagement. Your pack has a mass M and you're all perfectly bound together. Their pack has a puny mass, call it m. As soon as the ref says 'crouch, touch, pause, engage', your team surges forward at ramming speed V while their pack scrums down at speed v. This interlocking of heads, engagement, is governed, at least approximately, by conservation of momentum: the total momentum is the same before and after the packs intertwine. As momentum is the product of mass and velocity, your momentum before engagement is MV and theirs is mv. You'll push them back if MV is bigger than mv. So, a weighty pack that engages quickly will move the scrum forward just as the ball goes in. Too soon, of course, and the referee blows the whistle. Timed just right, life is much sweeter for your hooker. So, you need to have a heavy pack (to increase M) and put some 'oomph' into engagement to maximise V. Conservation of momentum is the key to getting quality ball from the scrum: the ball goes in as the pack goes down, the hooker scoops the ball back, and your greater momentum drives the opposition off the ball. You need never have one heeled against the head again.

Newton's second law, if we're dealing with something whose mass remains constant, can be simplified. When a force F acts on a mass m, it produces an acceleration $a = F/m$. The acceleration is how the velocity changes in time so it, too, is a vector. Its direction is the same as that of the force. Just like 55BC versus 1066 in British history, $F=ma$ is probably the second most-famous equation in the world – beaten out only by Einstein's $E = mc^2$.

Newton's second law is an extremely rich source of physics and contains a great deal of information. For example, m is the inertial mass of an object, which measures the amount of 'stuff' of which something is made. The force, though, is something entirely external to the object itself. Acceleration is a response of the object to this external force. Entirely different forces, be they gravitational, electric, magnetic, or whatever will produce the same acceleration, provided the forces all have the same magnitude and direction. Why this is the case is not obvious, neither to professional physicists nor philosophers[1]. The law governs almost everything we see in the world around us: bridges, rivers, dams, and almost all other works of modern engineering are based on this simple formula. It is part of what Nobel physicist Eugene Wigner described as the 'unreasonable effectiveness of mathematics in the natural sciences.'

The main aim of any pack is to drive the enemy backwards. From Newton's perspective, and also the referee's, the scrum should be stationary when the ball goes in. To get the Springbok pack to go back, the Aussies need to exert a greater force. This net force changes the scrum's momentum, and a change in momentum means a change in the scrum's velocity. The speed of the scrum goes from 0 to a new speed; the direction of the velocity, if everyone is square on in the scrum, is straight back towards the Springbok's line. Newton's second law tells us, unsurprisingly, that if you outpush your opponents, you go forwards and they'll go backwards.

We can do more, if we take some liberties with mathematics. In the ideal world of physics, we can imagine a pack where every player is the same mass, m. The pack's mass is then $8m$, and the scrum's mass is $16m$. Oddly enough, the idea that every forward has the same mass is not too far fetched. The Lions 2005 tour of New Zealand was a disaster in terms of results for the Brits – they were well and truly stuffed by the All Blacks. In terms of our physics approximation, though, the heaviest pack the Lions could turn out was pretty good. The total mass of this pack was 917 kilos, so that the average mass of a player was 917/8 = 114 kilos. If every player had the same mass, then each of them would have a mass of 114kg, but they didn't. The most massive packman on the tour, England second row Simon Shaw, was 121 kg; the least massive players – an honor shared by England flanker Lewis Moody and Scotland flanker Simon Taylor – were 105 kg each. Simon was 7kg more than the average mass, which is about 6% higher than the average. Moody and Lewis were 9kg lighter than the average, about 8% lower. Overall, the 'all players have the same mass' is not too bad an assumption.

Kilograms measure mass, not weight. Weight, measured in Newtons (whose symbol is N), is what you get when you multiply mass by the acceleration due to gravity. This acceleration, called g, is roughly 9.8 m/s^2 on the surface of the Earth, but changes slightly depending on where you are. It would be only one sixth of that on the Moon. Your weight can change by going to a different planet, being in a rapidly ascending or descending elevator, and a host of other ways. Your mass, though, remains the same, for it measures the amount of 'stuff', molecules, of which you consist. Mass is a scalar, but weight – a force - is a vector. Its direction is usually straight down towards the centre of the Earth.

In our perfect pack, if each player pushes with a force equal to his own weight, the forward push will be $8mg$. This, too, is not a bad assumption: A world-record weight lifter hoists barbells twice his own weight. We can expect a hard-working pack to make at least half that effort![2] If m is about 100 kg, then the pack pushes forward with a force of about 8,000 N.

How good is your hooker? Sometimes a bad one can work to your advantage. Suppose you have the put in. Your hooker won't push, for he has to heel the ball. The opposing hooker might fancy his chances of getting the ball, so he won't push either. That leaves 7 players per team pushing, and the honours should be even. On their put in, though, don't let your hooker go after the ball if he has fat chance of winning it. Instead, order him to push. In that case, your pack exerts a force of 8 *mg* whereas their pack musters a mere *7mg* of force. You therefore outgun them by the amount *mg*. The acceleration of the scrum is the net force (*mg*) divided by scrum's mass (*16m*), which is *g/16*. The scrum's acceleration, if we put in the numbers, is 0.62 m/s^2.

Acceleration is how velocity changes with time. The scrum, if the referee does his job, is originally stationary. The ball goes in, their hooker strikes, but within 1 second, the scrum is moving backwards at a speed of 0.6 m/s. In that one second, Newton's laws[3] tell us that the scrum has moved 0.31 metres, or about one foot. After two seconds, though, their pack is moving back at 1.24 m/s while the scrum has moved 1.24 metres backwards. Their pack is moving backwards at walking pace and you may even be able to push them off the ball. Newton's laws reveal the power of the eight-man shove!

Isaac also issues a stark warning and a challenge. The challenge is to get the pack in shape with some weight training. If each of your players exert $1/8^{th}$ more force than their opposite number, your pack will have, more or less, a one-man advantage. Newton's warning is that you can be in a sorry state if your star forward gets 10 minutes in the penalty box to mull over his 'conduct unbecoming'. Had he kept his temper, the pushover try would never have happened.

Seen a different way, what holds true for scrums also holds for rucks and mauls. The most effect way of winning phase two ball is to get bodies in there, pushing hard. In the 2008 Welsh tour of South Africa, in the Second Test the ball went to ground just inside the Welsh half but was coming back towards the Boks. The Welsh forwards, though, were there in sufficient numbers to spoil things and, in the broken play that followed, the ball was pounced on by Shane Williams. To score, Williams had to run half the length of the pitch and beat four South African players: within six seconds, Cymru had five more points. It was another case of severe Shane damage.

This model for pushing in a scrum, ruck, or maul is not particularly sophisticated. It does, however, give answers that are entirely believable. In international games, it's rare that a pack can make the opposition go backwards. Extra force, though, can take its toll. In 2006, England's pack consisted of a bunch of bruisers. In the game against Wales, the lighter, more mobile Welsh pack held its own for the first half and a large portion of the second. Then Martin Jones got sent to the sin bin, for an offence that was clear to the referee but not to anyone in the crowd. In Jones, M.'s absence, Wales still matched England in the scrum, but the extra effort the boys had to produce to keep things level drained their strength. Soon, England's pack ran rampant, and turned a previously close game into an undeserved rout.

In high-school rugby, some school teams can be entirely outgunned. I've been in scrums where, by the time the ball has come out, we've been going forward at running speed. In one game, we managed a pushover try from a scrum that started off at our opponent's 10 metre line. To be honest, I've also been in a pack where we've been pushed back at running speed, but it's not something I like to remember. Moving swiftly forwards or backwards is never pleasant if you're in the front row: the risk of someone slipping and the scrum collapsing is far too great.

The key to scrumptious scrums is to generate the greatest force that you can: The bigger the force, the larger the acceleration. Every player in the scrum exerts a force. The larger this force, the better your scrummaging will be. With apologies to my brothers in the pack, you'll have to suffer in the gym with weight training. The back, shoulders, and thighs that do – or transmit – the pushing need to be as strong as you can make them.

Weight training can get every pack member to increase the *size* of the force they exert. To maximise the push of the entire pack, though, the *direction* also needs work. The direction is important because force is a vector. If your pack is nice and level, then all of your effort is directed straight back towards your opponent's goal line. Seen side on, you need to 'align the spine' of the number eight and the second row. Then the power house of the scrum pushes forwards perfectly - entirely horizontal with the hips and backs of all the players at precisely the same height, directing all their effort straight towards the other team's tryline.

There's another advantage in keeping your back straight when pushing. Suppose your pack pushes slightly forwards, but downwards as well. If their pack does likewise, they exert a downwards force as well and, right where the necks of the front row meets, two downward force components combine. If these components are too great, the scrum may well collapse, risking a penalty or, far worse, serious injury for one of the front row players.

Isaac Newton tells the pack that good binding is vital, especially in the tight five. Make sure the locks are square on to the scrum. That way, they won't drain vital energy pushing at a slight angle while the number eight pushes at a different one. A pack that binds together, which pushes parallel to the ground and square on to the opposition, maximises the force it exerts. The South African pack may be huge (the current one is about 900 kg), but if England (currently 865 kg) has superior technique, the Springboks will go backwards. So, another great way to maximise the force is to make sure your pack is square on. That means spending hours on the scrummage machine.

Some people stick out in a crowd. While six players push forward, two do not. Those flankers, wing forwards, call them what you will are usually at some angle or other to the main scrum. Force, lest we forget, is a vector, with size and direction. Everyone else pushes straight down the pitch, but the flankers are at an angle. In terms of mathematics, we can split the force exerted by a flanker into two directions, called components. The first component, along with everyone else, is down the pitch. The second component is sideways, across the pitch. In other words, the two flankers expend part of their energy moving the pack forward, and another part squeezing the pack together. If the flanker is at an angle of 45 degrees, the two components of the force are equal in size.

Isaac Newton's laws of motion dispense free rugby advice. If you're having a hard day in the scrums and things aren't working for you, what can you do? The angle at which flankers bind on can be changed easily. This angle determines how much of their push goes downfield and how much goes across field (see Figure 1.1). If the binding angle is zero, your flanker pushes downfield with all his might. If he binds at an angle of 30, 45, or 60 degrees, then his force downfield is respectively 87%, 71%, and 50% of the total force he can exert[4]. Thanks to physics, the strategy is clear: On their put in, get your flankers to push at a far smaller angle than they have been doing, and get the hooker to push as well. That will maximise your forward force.

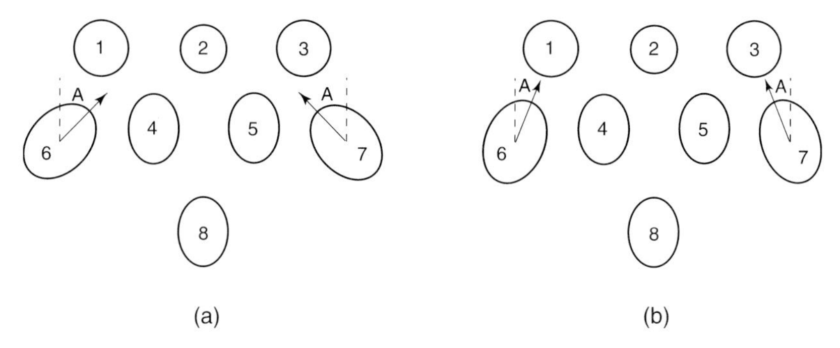

Figure 1.1: Flankers at an angle.
The binding angle A determines how much force is directed downfield:
(a) steep angle, small force (b) shallow angle, large force.

Suppose their flankers are at 45 degrees and their hooker doesn't push. If all three players push with a force F, their combined efforts amount to a total force towards you of $1.41F$. Your flankers, though, are straight on and your hooker pushes. Your trio exerts a force of $3F$. This gives you an advantage of about $1.6F$. If the hooker and two flankers can all push their own weight, and each player's mass is 100 kg, then your team has an advantage of 1,500N. That may be enough to halt their pack's progress.

If your pre-season training has paid off, so you win scrums easily, then Newton again shows you what to do. If you're having great success, put your flankers at a wider angle. They will contribute far less to the forward force your pack exerts, but that's ok — you have plenty of extra force to play with, otherwise the enemy wouldn't be going backwards. Steeply angled flankers compel the opponent's scrum half to go the long way round in his attempt to swipe the ball. You still win the scrum and you still push them backwards, but now there's an added bonus: you protect your scrum half by forcing theirs to take the scenic route around the wing forward. This gives your number nine more time to feed the ball. The last benefit, always a joy, is that with a large flanker blocking his view of the ball, it's much easier for their scrum half to blunder offside. Remember to thank Isaac Newton for the points as your place kicker sends the resulting penalty between the posts and over the bar.

All of these ideas are put into practice when the game is played at a high level. If you ever watch videos of great flankers in action, such as Ireland's Fergus Slattery, or France's Jean-Pierre Rives, you'll see how they changed the angle at which they bind, depending on how tough things were for Ireland or Les Bleus in the scrum. And, of course, whether they thought they could hammer the outside half this time…

Another side effect of Newton's laws, if you're having a good day, is that your hooker doesn't need to push when it's the other team's put in. If you're roughing them up in the scrums, let your hooker have fun trying to embarrass his opposite number. It's a great feeling to hook a couple against the head. Likewise, once you've got them going backwards, you have the luxury of exploring with number eight pickups or even letting your flankers run with the ball. Remember, though, that the second they break off, the flankers cease to exert a force. The force your pack exerts is reduced, which means that the speed at which the scrum is moving will either be lessened or, horror, reversed. In addition, the mass of the pack is less than it used to be: it's gone from, say, *16m* down to a mere *14m*. You will have lowered your pack's force, making it easier for their pack to push you backwards, *and* you've reduced the mass of the pack, which will increase the pack's acceleration - towards your own goal line. Try to remember this when play is dangerously close to your own try line. Of course you want to break off from the ruck or maul, to make sure that you have time to spot which back has been fed the ball so you can hammer him before he can touch the ball down. Break off too soon and the ghost of Newton will point out that you have reduced your team's force and decreased the mass of the maul. Don't be surprised to see your opponents keep the ball in the maul and drive forward. Instead of bathing in glory for the try-saving tackle, you get to watch their forwards trundle over the line and score, right at your feet.

I used to play on a team where the Number 8's nickname was 'Tank', for that's what he was built like. A number-8 pickup was never an option for us, because without Tank's effort our pack had no push. On the few occasions when he did pick up the ball, we usually went from holding our own to being pushed backwards rapidly. Tank was not the most rapid sprinter on the face of the planet, so sometimes he would pick the ball up, only to get tangled in the legs of our rapidly retreating second row. It was embarrassing – and dangerous.

One of my favourite number eights was Andy Ripley, once of Rosslyn Park and England. (For the record, red-and-white hooped Rosslyn Park has the honour to have played in the first international rugby match on Continental Europe. On April 18th, 1882 they defeated Stade Français in Paris, four converted tries to nil). Returning to Ripley, he was a strong, fearless runner -- the kind of back-row forward you'd love to see pick up the ball and run straight back at the opposition. I wouldn't have liked to try and stop him. Ripley went on to become Britain's Indoor Rowing Champion. Indoor rowing, on an ergometer, is one of the most hellish sports on the face of the planet. It requires incredibly strong thigh muscles, for most of the force a rower exerts comes from the quadriceps muscles. These are the same muscles that you use to lift your legs when running, or when pushing. It's no wonder that Ripley excelled on the erg, where he still holds two world records, as well as on the pitch. He also excelled, though he may not have realised it, in putting Newton's laws to their best possible uses.

LOCK SOLID: NEWTON'S THIRD LAW

The last of Newton's three laws is that action and reaction are equal and opposite. Every player has a certain mass, and gravity causes each player to have a certain weight. Our weight force pushes us down towards the centre of the Earth. Luckily, because action and reaction are equal and opposite, the Earth pushes back an equal and opposite amount. If it didn't, we'd sink into the ground. Now, if you happen to weigh 75 kilos, you may have noticed that some other people, who weigh 175 kilos, still don't crash through the Earth's crust. The Earth can easily supply a reaction force of about 1750 Newtons. This has its application in the scrum, too. If your pack is outgunned, stop pushing! Your pack should dig in, lock legs, and keep low and straight. In so doing, you hope to transmit all of the force exerted by the other pack down into Mother Earth, who can easily provide a reaction that is equal and opposite. Your pack may not be as strong as theirs, but you can get the Earth, with a mass of more than 10^{24} kilos, to play for your team.

WHEELING THE SCRUMS, ROLLING MAULS: A TRIUMPH OF TORQUE

There will always be games where your pack is going to have a difficult day. Part of this can be psychological: teams who come out of the tunnel wearing broad hoops will usually look shorter, but beefier, than those who wear plain jerseys. So, especially in high school, remember that if the other pack looks short and stocky, it might just be a trick of the eye.

If you do get beaten up in the set piece, though, try your best to throw a spanner in the works: wheel their scrum. A spanner is a pretty good model of what you need to do. If you've ever tried to get a rusty nut undone, you know how much effort it puts on your fingers. Put a spanner around the nut and you can apply far less effort but get the nut to move more easily. The difference is torque. Torque turns, the same way that forces push or pull.

If you push with a force F perpendicular to the end of a spanner of length L, then the size of the torque you produce is FL. Torque, too, is a vector. Its direction is perpendicular to the force and the spanner. This direction is either into the nut (clockwise) or out of the nut (anticlockwise).

A force pushes things forwards or backwards, left or right. A torque turns things in a circle. To get the nut undone – to get it to move in a circle so that it unscrews – you need to apply a certain torque. You can apply a whoppingly large force with your fingers (because L is close to zero) or a relatively small force a long distance from the nut. Hence the spanner – a simple tool that lets you increase that distance (See Figure 1.2).

To wheel the scrum, you need to exert a torque. Instead of the nut, you want the scrum to rotate. The point of rotation is not the nut, but the hooker's head – though some hookers can be really hard nuts indeed. Unlike a spanner and a nut, the scrum is more like a seesaw. If you have two kids of equal weight sitting at opposite ends of the seesaw, nothing happens. The torque is balanced. If a beefy prop sits at the high end of a seesaw (just try

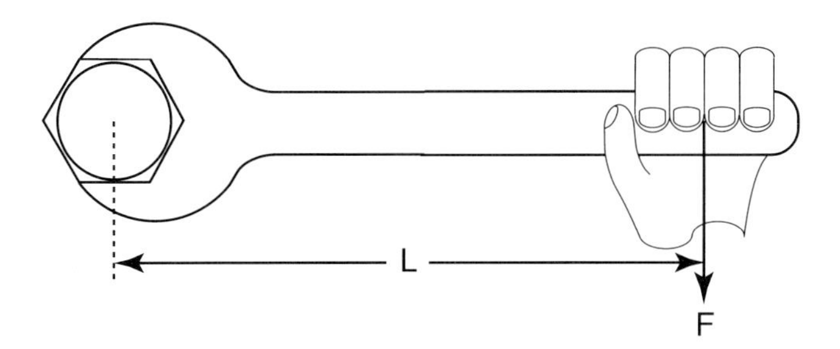

Figure 1.2: A force F at a distance L from the nut generates the torque FL.

getting Phil Vickery to do that!) and a spindly winger on the low end, the seesaw will rotate – the distance from the balancing point is the same for both players, but the forces (their weights) are not equal. This causes a net torque, which makes the seesaw rotate about its middle balancing point, known as the fulcrum, jettisoning the winger into the air. The seesaw and the spanner are two simple tools, but both of them are levers. As Archimedes said (the Ancient Greek scientist who, like many rugby players, ran through the streets naked shouting 'Eureka', Ancient Greek for 'I've found it') 'with a lever I will move the whole world.' We don't need to move the world, just your opponent's pack.

To unbalance the forces in the scrum and so create a torque, play the angles. Suppose it's their put in at the scrum and they're close to your line. Take a chance and guess that, unless you heel against the head, they'll want to bring the ball out on the open side. So, get the open-field boys in your scrum to push mightily, while the blind side goes limp. In contrast, your opponents push with equal force to the left and right of your hooker's nose. *They* exert no torque at all, for their force is balanced. Only one half of *your* pack pushes, so that produces a torque. The greater the torque, the more swiftly the scrum wheels.

To get a rough idea of the torque, suppose every player on the right-hand side of the scrum exerts a force F and every player is of width W. (See Figure 1.3). Your hooker and number 8 exert no torque, for all of their push goes straight through the fulcrum, so that their value of L is zero.

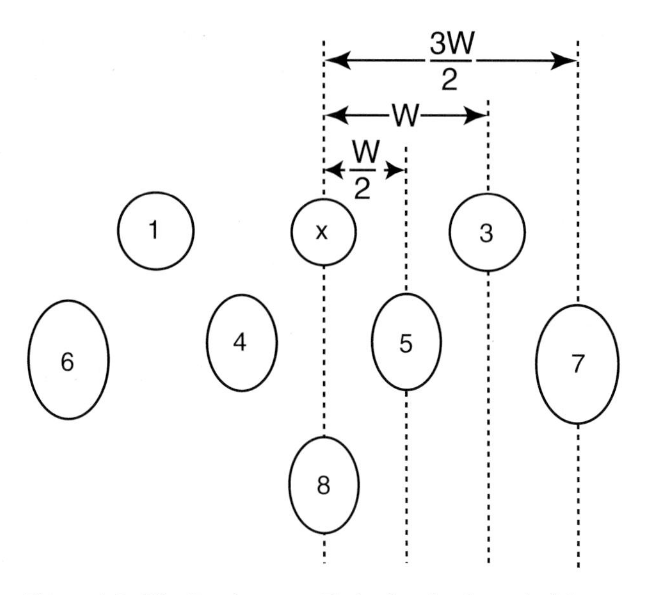

Figure 1.3: Wheeling the scrum. The hooker's head, x, is the fulcrum.

The lock who pushes is about half a body width from the fulcrum, horizontally speaking, so generates a torque *FW/2*. The prop is a whole body width, creating torque *FW*. The flanker who pushes is about 1 ½ body widths, *3W/2*, from the fulcrum and so throws in some torque in the amount of *3FW/2*. Doing our sums, we've now got a total torque of about *3FW*. If each player is about two metres tall, then their wingspan – from fingertip to fingertip of outstretched arm – is about 2 metres as well. So, we can guesstimate a player width, *W*, of about 2/3 of a metre. The total torque tormenting their pack is about *2mg* – assuming, as usual, that each player exerts his own weight and that all players weigh the same. If our players average 100kg, then this torque is about 1,960 Nm. But not everyone is equal. One half of all the torque is generated by just one player, the flanker, not because he's beefy and brawny, but because he's the greatest distant from the hooker's nose. Mind you, if he is beefy and brawny, he can exert a greater force. The lesson is clear: if you have one solid flanker and one spindly one, get the solid flanker to do the pushing needed to generate the torque that will wheel the scrum.

Torque turns; that's its job. The important question is how quickly a certain torque can wheel the scrum. The answer is not simple. Unlike a force, where all we do is add up the mass to work out the acceleration, torque requires something more complex. How something responds to a torque depends not only on the mass of that something, but also the way in which that mass is distributed. It is summed up by the quantity known as the moment of inertia. As a rough estimate, if a conker of mass m whizzes around on a string of length L, the conker has a moment of inertia mL^2.

Working out precisely the moment of inertia for a large herd of forwards is a thankless task. Imagine having to measure the mass, length of torso, and shoulder width of every member of the scrum – even if they'd let you. Life is too short. Instead, a physicist will create a model -- an educated guess.

If you squint hard enough, the scrum looks like a solid block of flesh. Call the length of this block, from the studs of one number 8's boot to those of his counterpart, A. The width of the scrum, from the left shoulder of the loose-head to the right shoulder of the tight head prop, is B. We can now relax for mathematicians, bless them, have long since known what the moment of inertia of a box rotating about its midpoint is. Physicists, fortunately, can use the hard-earned knowledge of mathematicians by looking up the moment of inertia in a book, or on the web, rather than laboriously having to calculate it. The answer is $M\left(A^2+B^2\right)/12$.

Putting in some numbers, the shoulder-to-shoulder width of the front row, B, is a cosy 4 metres, say. For A, the distance from the toes of Scottish number 8 Allister Hogg to those of Italy's Sergio Parisse, 7 metres looks reasonable. For England against South Africa, with the packs of 865kg and 900 kg respectively, the mass M is about 1,765 kg. Sticking all these numbers into the formula suggests that the moment of inertia of the scrum is around 9,500 kg m^2. (More precisely, it's 9,581 kg m^2, but as were guessing the size of quantities, 9,500 is good enough.)

The torque, we estimated earlier, is about $3FW$. For the South African pack of 2008, this torque would be about 3x (900/8) x 9.8 x (2/3), which is about 2,200 Nm. Just as a force acting on a mass produces an acceleration, a torque acting on an object generates an angular acceleration. To find the angular acceleration, divide the torque by the moment of inertia. The Lions' pack when wheeling could generate an angular acceleration of 2,200/9,500 radians/s^2. This is 0.23 radians/s^2, slightly above 13 degrees/s^2. After two seconds, the pack will have turned through an angle of about 26 degrees.

This has truly started to mess up their scrum. If the ball was put in on the five metre line, their scrum half now has to pick up the ball about 3 metres from the touchline. If your blind side flanker is alert, there will be little chance of the scrum half sneaking past him. The only option is to go to the open-field side, where your ever-alert backs stand waiting in battle formation, flat, eagerly expecting to crush their opposite numbers. Of course, if you can hold on a bit longer, it takes about 3 seconds for the pack to spin a full 90 degrees. A bit longer and (law 20, paragraph 18 to the ready) the ref sounds the whistle and the scrum must reform.

The answer is pleasing. Physics dictates that you can wheel a pack in a few seconds – not bad, seeing as we used an unsophisticated model. We could refine it to get a more realistic answer, but that's not worth the effort[6].

The important lesson is that the more the force, the greater the torque, and so the less time you need to wheel the scrum. The scrum wheels more quickly if it is lighter, so getting some of your players – I suggest at least the blind-side flanker -- to break off once the ball goes in could be of great help.

All in all, physics and reality coincide. Our estimates for scrumming show that it's pretty hard to push another pack off of the ball. It's fairly simple, though, to arrange your flankers and other pack members to create a torque on the pack. That's probably why, when watching high-standard rugby, you're far more likely to see scrums wheel, rather than one side stuffing the other.

The antidote to being pushed off the ball, or having your scrums wheeled mercilessly by the opposition, requires complex physics. There is a far simpler solution, though – make sure your prop is 'injured.' Without a back-up prop on the bench, the referee will insist on a static scrum, where no-one is allowed to push. Unsporting? Possibly, but it's all part of the game.

Lineout leaps

It's great to dominate the scrum. It also pays to rule that other set piece, the lineout. England ended up with a comfortable win over Italy in the 2006 Six Nations tournament, but the victory was not as easy as the 16-31 scoreline suggested. The main cause of victory was England's total dominance over the Azzurri in the lineouts. What the Italians lacked was a great jumper.

The world's greatest jumper is not a human being. It's an insect called the froghopper, a.k.a., the spittle bug. This mighty insect can launch itself about 70 cm into the air, even though it's a mere 6 millimetres long. Put another way, it can jump 700/6 = 117 times its own height. If England's Steve Borthwick or Ireland's Donncha O'Callaghan (both 1.98 m) could jump 117 times their own height, they could catch a ball thrown 234 metres into the air – about the height of the Canary Wharf tower in the heart of London's renovated Docklands region or, for those headed to the Parc des Princes to watch Les Bleus, two thirds the height of la Tour Eiffel.

The froghopper uses a cunning catapult device to throw itself so high. A second row, however lofty, has to rely on brute strength and a couple of colleagues to hoist him high. The important thing for any lineout jumper is to maximise the vertical leap. Start off standing by a wall, raise your hands and make a mark on the wall with your outstretched fingers. (Hang a sheet of paper over the wall and use finger paints, if you feel so inclined). Then jump as high as you can, from standing, and mark the wall as you reach the maximum height of your jump. The distance between the two marks is your vertical leap.

Beware of false advertising. Basketball great Michael Jordan, formerly of the Chicago Bulls, has a 'vertical leap' of some 1.2 metres (48 inches), but this is after a suitable run up, not from standing. He is dwarfed, though, by French Algerian hoopster Kadour Ziani, whose record stands at 1.4 metres (56 inches) – though again he has to run up.

A vertical leap from standing of about one metre is world class, and the records are usually held by weight lifters, not basketball players or high jumpers. This is the hint: To improve your lineout jumping, get bigger and more flexible hips, just as weight lifters do. Flexible hips give you a springlike, or catapult launch, rather like the froghopper. Stretching exercises and yoga will help with hip flexibility; weightlifting helps develop powerful thigh muscles needed to get you off the ground. The flexibility may also help you to survive if a scrum or maul collapses and your legs are at an odd angle. Flexibility may help you avoid the ominous 'crack.'

Newton's second law says that if you blast off a speed v, you'll reach a maximum height of $v^2/2g$. So, to produce a vertical leap of one metre, you need a launch speed of about 4.5 metres per second, which is a fair clip. The important point is to have an explosive launch: If you double your jumping speed, you'll quadruple the height you reach.

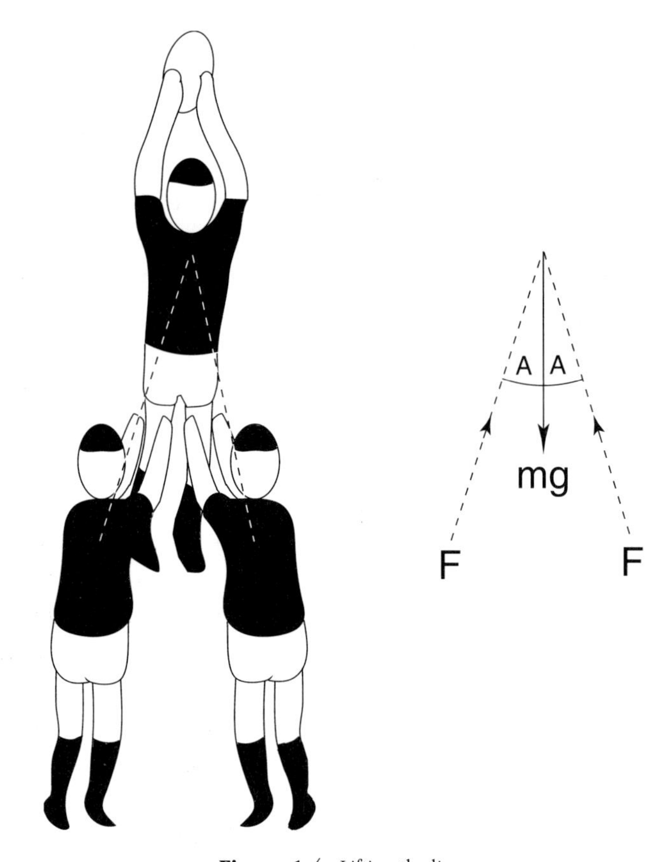

Figure 1.4: Lifting the line.

A couple of sturdy yeomen from the front row can help to lift you (see Figure 1.4). They, believe it or not, generate torque. Their shoulders are the pivot, and the force is applied at the end of their arms. At the moment of takeoff, their arms are horizontal and so no force goes to helping you jump. As you go higher, though, the angle made by their arms swings from zero degrees and will eventually get close to 90 degrees. In that case, when you're at the highest point of your jump, the vertical force supplied by your lifters is at maximum industrial strength. Instead of gravity being the only force in town, your new force will be *2F –mg*, where *F* is the force exerted by one of the gentlemen of the front row. As we've always assumed you can push your own weight and that all pack members weigh the same, you now soar like an eagle, for the net force acting on you is *mg*, and it's upwards. If you and the two props are 2 metres tall and have arms of length 1 metre, then you can catch a ball thrown 5 metres high, assuming the props hoist you by your middle. (Two metres is not bad; the RBS Six Nations web site gives the honours for tallest players to Wales's Brent Cockbain and Ireland's Matthew O'Kelly, both of whom measure 2.04 metres). For those having trouble sleeping at night, forget the jumping sheep and turn your thoughts to the 100+ page rule book for rugby. In case you missed that section, rule 23(15)(c) says that you can't be hoisted before the hooker has thrown the ball, the dire consequence being a free kick to the opposition 'fifteen metres from the touch line along the line-of-touch.'

The hooker can throw the ball in a beautiful arc that peaks five metres above the turf (two metres for you, one metre for your extended arms, and two metres from your lifters). It's wonderful physics – unless the touch judge indicates the throw wasn't straight. Our equation predicts that the lighter the person jumping and the stronger the people lifting, the higher you'll go. Here's a thought: get the lightest flanker to jump and have the second row do the hoisting. It might be worth a try – literally.

Have some sympathy for the poor hooker, though. If your supporters push you up with a force of *mg*, you will be raised to your maximum extent above the ground (5 metres) in about three quarters of a second. The number two has to throw the ball in so that it is exactly 5 metres high after this 0.75 seconds – no matter whether you jump at two, four, or six in the lineout. If the hooker lets go of the ball when it's two metres above the ground (so he throws from about head height), then Newton's second law predicts[7] that if the ball hurtles from the hooker's hands at speed v, it reaches a maximum height of 3 metres a mere 0.75 seconds after travelling a distance x only if the launch angle has a tangent equal to $5.625/x$. If the jumper is second in the line, then he's about 6 metres from the hooker, so that tan A is about 0.9375, which means the launch angle is about 43 degrees. Isaac also demands that ball speed obeys $v = 4x/(3\cos A)$. In other words, to reach the number 2 jumper, the hooker has to throw the ball at 11 m/s. If there's a different signal, the ball has to reach the number four man, who is about 8 metres in from the hooker. In that case, the ball has to be launched at 35 degrees with a speed of 13 m/s. Last, if the jumper is at number 6 and thus some 10 metres away, the ball has to be hurled in at 35 degrees, with a speed of 15 m/s.

These are fairly precise, so your hooker has a demanding role. A couple of metres per second off on the launch speed or five degrees wrong on the launch angle and what was meant for the number 2 jumper ends up careening over the full lineout, entering the outstretched, surprised, yet grateful arms of their inside centre. If that were not enough, the hooker has to throw the ball in having sprinted for most of the eighty minutes, even having blurred vision from a knock on the nose. Perhaps it's no wonder that the website 'England Rugby' proclaims the hooker as 'a key team member', who – together with the fullback, fly half, scrum half, and number eight – forms the 'spine of the team.' In the past, Gloucester, England, and the British and Irish Lions have given the number two shirt to a player described as the fittest person in rugby union -- Andy Titterall. I share with Andy the distinction of having gone to the Hugh Christie School, but sadly I share neither his fitness nor playing ability.

Mauls: The manifestation of momentum

While I enjoyed scrums and tolerated lineouts, I loved mauls. There's nothing quite so fine as going into a melee, ripping a couple of hands off the ball, and working the ball back to the scrum half. For the England team of the late seventies, mauling was a way of life. Charles 'Crashball' Kent had been on the Oxford rugby team for five years, during his training as a medical doctor. He went on to play for England and the Barbarians as the inside centre. During his brief international career – he was once uncharitably selected in the honour roll for 'England's Worst Fifteen' – I never once saw him pass. This probably wasn't his fault, for in those days the England backs could barely pass twice before dropping the ball. With such a large English pack, Kent's job was to run straight until stopped. Then the maul would begin and, by sheer weight, the white of England (chosen to commemorate the colour worn at Rugby School, where the game was founded) would move forward. Unless the opposition could steal the ball away, it would be England's put in at the scrum, which Peter Wheeler would inevitably win. Then, the process would be repeated.

If you arrive early at the scene and the ball is visible, go for it. If you are a late arrival on the maul scene — and why are you? — there's more advice that can be garnered from Newton's laws of motion. If no external force is present, momentum is conserved[8]. Not 'conservation of the rain forest' type of conservation; physicists mean that a certain quantity remains constant. For momentum, a vector, we mean that the size and direction remains the same before, during, and after an event. This simple concept is the basis for a branch of physics known as collision theory. This describes the realm of subatomic physics, where particles are accelerated to phenomenal speeds and smashed into each other, in the hopes that rare and exotic particles can be created, ones whose properties are described by the quantum mechanical, relativistic form of collision theory. Right now, scientists are at work on the Large Hadron Collider, a device they hope will lead them to discover the Higgs boson, the 'missing link' of particle theory. The same laws also apply to everyday collisions, such as car crashes.

For those playing rugby, collision theory means something far more important. Suppose a maul has already formed, one that is stationary. The initial momentum of the maul is zero. If a 'particle' — a forward — of mass m runs at speed v, the total 'maul plus player' momentum before the collision is a mere mv. Once the forward crashes into the maul, he is absorbed by the big blob of mass M. After you've joined, the new maul has a total mass of $M+m$. As the momentum before and after has to balance, the enlarged maul now moves with a speed V, which has to equal $mv/(M+m)$, or else the momentum bookkeeping wouldn't add up. This is only a model, remember, so we've simplified things greatly. The players are now hard, tiny particles smashing into a blob, the maul, which is another hard, tiny particle but of higher mass. (In real life, the maul is made up of beefy people who can change their direction without needing to collide off the maul or another player.)

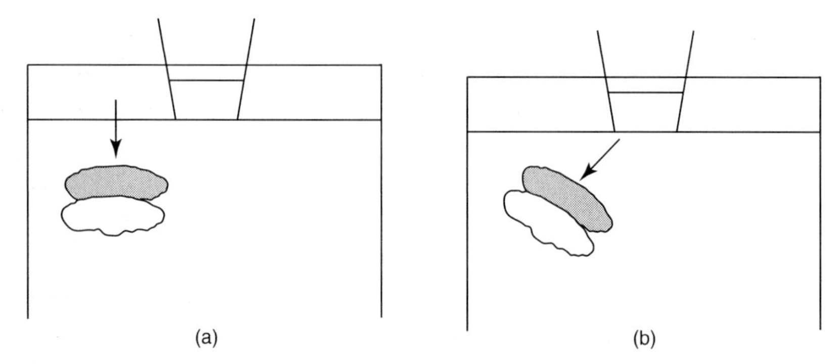

Figure 1.5: Close to your try line (a) smack into the maul straight ahead if the ball is coming out your side (b) or towards touch if it's on their side.

Still, our stripped-down model suggests you should add a couple of items to your to-do list. First, crunch into the maul at the highest speed you can muster. The higher the speed at which you join, the greater the speed the maul moves at afterwards. Don't forget, all things being equal, if the ball gets held up in the maul, it's the side moving forward that gets to control the scrum. Second, the bigger the maul, the larger M will be, so the smaller the effect becomes. If the maul starts off with two players on each side, the next player to arrive at speed V makes the maul move at $V/5$. If the stationary maul has 7 players on each side, the late comer will make it move at a speed of $V/15$.

That's the joy of having the Hair Bear props, Duncan and Adam Rhys Jones, on your team. These shaggy stars of the Ospreys and Wales not only rank among the world's largest props, they also rank among the fastest. Adam is 127kg while Duncan tips the scales at 110kg. This hirsute duo can run at speeds getting close to 10 m/s. If they smack into a stationary maul of, say, four 100kg players, they'll shove it forward at a speed of about 3.75 m/s. There's lovely!

Late-arriving, slightly unfit players know you can lean gently on the maul without anyone really complaining. There's no way for the coach to check whether you're pushing hard or taking a breather, for the well-formed maul won't move much. Even so, you should always hit an open-field maul at ramming speed. It may not gain you much territory, but you may earn the put in at the scrum.

There is an alternate plan. Suppose the maul is close to the corner flag. If you're defending and the ball is going to emerge on *your* side, then you probably ought to smack into the maul at ramming speed, heading straight up the field. This imparts maximum linear momentum to the maul, which should give your scrum half more room to work with. If the ball is about to come out of the maul on your *opponent's* side, you might want to change your angle of attack, trying to knock the maul, and thus the ball, into touch. (See Figure 1.5). A line-out close to the try line is always a concern, but it's better than having the opposing backs sprinting towards you from less than five metres out. Which would you sooner face – an Australian lineout or Mat Rogers running towards you at full speed from five metres? I'd choose the lineout.

If you're attacking, you have some choices. If your team expects to get the ball, then slamming into the maul at top speed gets you closer to the try line. You could try another tactic, which is to hit the maul hard at the edge, giving it maximum torque. This causes the ball to rotate, allowing your team to form a rolling maul and keep the ball moving, until it's time to recycle it to the scrum half.

Maximum torque is also a good idea if they look as though they'll get the ball a few yards in front of their try line and a few feet from touch, for you have now messed up the scrum half's options. Your impact, literally, has knocked the maul closer to the try line and rotated it over towards the touchline. The scrum half now has hardly any room to play with. What's more, as you joined the maul close to its fringes on the open-field side, keep your eyes peeled. The scrum half is cornered; he has no good angle to kick the ball and probably doesn't want to trundle into touch. His preference may be to risk the pass to the outside half. You, on the open-field side of the maul, can anticipate the pass, and nail the fly half just as he receives it – especially if the ref has a poor track record in spotting forwards who encroach offside. Good hunting!

CHAPTER ONE - ENDNOTES

1 For constant mass, the force from Newton's second law is $\vec{F} = m\vec{a}$. Here, m is the *inertial* mass. For the case of gravitation, the magnitude of the force is $m_g g$, where m_g is the *gravitational* mass. It turns out that these two masses are, as far as we can tell, the same. There is no reason (that we yet understand) why this should be the case.

2 There's further evidence to back up our claim. A standard problem in physics is to find out what happens when you place a ladder against a wall. The wall pushes back on the ladder with a force that is fW, where f is the coefficient of friction between the ladder and the ground, and W is the weight of the ladder. Replace the ladder by a hefty, rigid, number eight and the wall by the derrieres of the second row. The coefficient of friction between rubber (the sole of the boot) and grass is about 0.4. So, the number eight just by leaning should be able to exert a forward force of $0.4W$, or about half his weight. The use of studs increases the force he can exert without slipping. So, a force of somewhere between ½ W and W is not unreasonable.

3 For a constant acceleration a from a scrum initially at rest, Newton's second law gives $v = at$ and $x = at^2/2$. Here v is the speed after t seconds, and x is the distance travelled.

4 If the angle between the pack and the flanker is A and the force exerted by the flanker is F, the component downfield is $F\cos A$ while the cross-field force is $F\sin A$.

5 Rather like exams in English classes, force is to acceleration what torque is to angular acceleration. So, if the torque is constant, the pack turns through an angle A in a time t where $A = (N/I)t^2/2$. Here N is the torque while I is the moment of inertia. (Compare with $x = at^2/2$).

6 One consequence of our torque equation is that increasing the torque by 10% will reduce the time taken to wheel the scrum through a given angle by about 5%. So, getting a more precise estimate of the moment of inertia of the scrum won't change our answer that much.

7 In the horizontal direction, the distance travelled in a time t by a ball launched at angle A at speed v is $x = (v\cos A)t$. The ball's height is $y = (v\sin A)t - \frac{1}{2}gt^2$. The maximum height the ball reaches occurs at time T, where $T = v\sin A/g$. But $T = x/(v\cos A)$. Combining these two equations gives $\tan A = gT^2/x$. If we put g = 10 and T = 0.75, this gives tan A = 5.625/x.

8 From the second law, the force acting on body number 1 caused by body number 2 is $\vec{F}(1,2) = d\vec{p}(1)/dt$. Likewise the force acting on body number 2 caused by number 1 is $\vec{F}(2,1) = d\vec{p}(2)/dt$. From Newton's third law, action and reaction are equal and opposite. In other words, $\vec{F}(1,2) + \vec{F}(2,1) = 0$. More explicitly: $\vec{p}(1) + \vec{p}(2) = m(1)\vec{v}(1) + m(2)\vec{v}(2) = const$, which is the equation of conservation of momentum.

Chapter Two

The Back Attack: On running, passing and scoring in the corner

'Remember that rugby is a team game; all 14 of you make sure you pass the ball to Jonah.'

Anonymous fax to the New Zealand team

'In 1823, William Webb Ellis first picked up the ball in his arms and ran with it. And for the next 156 years forwards have been trying to work out why.'

Sir Tasker Watkins VC GBE

One of the great joys in watching Five Nations rugby in the 1970s was seeing JPR Williams in action for Wales. With apologies to France's Serge Blanco and Scotland's Andy Irvine, for my money JPR was the best fullback ever to play the game. (He may not be the most famous person to occupy the position: Yale University once gave the task to future president George W. Bush). Williams was one of rugby's foremost sprinters and every time he burst into the line he threatened to score. Also, he had an amazing running motion; as he sprinted, his knees came up incredibly close to his chest. It was not merely an odd way to run: it was a devastating weapon. Seldom would a game go by without some hapless opponent trying to tackle JPR, only to have the Welsh wizard's kneecap crack into his skull at maximum speed. Fortunately, Williams was not only a fine rugby player, he was also a medical doctor. The opponent became JPR's patient, with no lines and no waiting, while Dr. Williams diagnosed whether the player was too badly concussed to stay in the game. JPR now works in a hospital as an orthopaedist near Bridgend, not too far from where he played his club rugby.

There have been other great sprinters to play the game of rugby, some of whom were Olympic class. (Lest we forget, Eric Liddell — whose life story is partly told in the movie *Chariots of Fire* — not only won the 1924 Olympic gold in the 400 metres and bronze in the 200 metres, he also played on the wing for Scotland). Physics can tell us more about such speedsters. Over the years, models have been developed to describe what goes on when an athlete runs as quickly as possible. Other models account for how long-distance runners move. For the backs, we need the mathematical model that is most appropriate for the fast and the furious.

RUNNING FOR DEAR LIFE

Isaac Newton's second law says that a force changes momentum. So, a player has to generate a force if he wants to move from rest to maximum speed. It doesn't particularly matter what constitutes the force; we say only that the muscles, sinews, and bones all combine to generate it. The big idea is to make things as simple as possible, but no simpler. As a rough guide, sprinters tend to be fairly solid individuals, whereas distance runners are usually thinner. There's a reason. The force a sprinter generates has got something to do with the drive of the arm and leg muscles, which are joined together by the pulling action of the abdominal muscles. Punch those arms, lift those thighs, and you might sprint as quickly as JPR.

Einstein claimed, in 1905, that nothing can travel faster than the speed of light in a vacuum. If we could run with a constant force, with nothing to slow us down, then in principle we could get close to the speed of light. In practice, there is always some resistance to motion. We know full well that there's a pretty definite maximum speed that we can sprint at, lurking at around the 10 m/s mark, so there has to be a fairly strong resistance to motion. The simplest way to model this is to make an educated guess. It seems reasonable that the faster you go, the greater the resistance becomes. If we believe that if you double the speed, you double the resistance, then we can say that the resistive force is some constant multiplied by the speed at which you run. For our basic model, we don't really care what the cause is. This is the way physics works: craft a simple model, see what you get, discard it if it doesn't describe reality well, or refine it if it does.

From Newton's second law, we know that the force exerted by the runner, once you subtract the resistive force, creates a net force. That's to say, whatever force is left after you've overcome resistance goes to produce acceleration, rather like whatever cash you have, after you've paid your bills, you can spend.

A specific version of this equation, suitable for fast runners, came from Archibald Hill. A physiologist by training, Hill won the 1922 Nobel Prize for his work on muscles, but he also served time as a member of parliament for Cambridge University, in the days when Oxford and Cambridge Universities had their own MPs. Hill asserted that a sprinter runs so that the force he or she exerts is constant. In other words, over 50, 100, 200, or 400 metre sprints, the runner exerts the same force from the start to the end of the race. Hill's model is that the force exerted by JPR in a sprint for the line is the maximum it can possibly be; call it F_{MAX} . That is, human beings can, for short bursts, exert a constant, maximum force.

Think about top speed. At some stage, you sprint as fast as you can and, try as you might, you can't go any faster. In other words, you are not accelerating, so that the maximum force you can exert, F_{MAX} merely equals the resistance force R to your motion. From Newton's law, we know:

$$ma = F - R$$

For sprinting, Archibald tells us that F is our constant, F_{MAX} , so:

$$ma = F_{MAX} - R$$

At some stage, the resistance balances the maximum force you can exert, and that determines your top speed, for then $a=0$.

To try and describe this, we use what physicists call a 'back of the envelope' model. We make the mathematics simple and attempt to get a big picture answer. It may be (and almost certainly will be) wrong in the details, but if the answer is plausible, we might have a physics-based description that's on the right track. We can then go back and beef up the mathematics.

Because the resistance depends on the speed, the net force F_{MAX} -R changes with time. To avoid this, we replace the actual force by an average force.

When JPR starts running, the resistance is zero, so the net force he exerts is F_{MAX} . When he sprints as fast as possible, F and R are equal. There's no acceleration because there is no net force. The average net force is the average of F_{MAX} and 0, which is $F_{MAX}/2$, and this is constant. We can now use Newton's laws for constant acceleration, which makes the maths much simpler. The constant acceleration is $F_{MAX}/2m$, so after a time *t*, JPR gallops along at a speed

$$v = \frac{F_{MAX}}{2m}t$$

Purists might worry about using an average force, but it's a bit like driving a car. You might go along some stretch of road at 25 km/h, another stretch at 35 km/h, but rather than worry about your speed at any particular moment in time – a mere detail – we can get the big picture by looking at your *average* driving speed, which is 30km/h, give or take. The police, lest we forget, will worry about your *instantaneous* speed. If you are caught doing 59 km/h in a 30 km/h zone, you can't get out of a speeding ticket by promising to drive at 1 km/h for the next hour…

If JPR's maximum speed is *u*, our model says he'll reach that speed in a time $T = 2mu/F_{MAX}$. We can now use our rule of thumb, that every human exerts a force equal to their own weight. In that case $F_{MAX} = mg$ and JPR is unstoppable after $2u/g$ seconds. Assuming he can run at about 10 m/s, we expect him to be up to top speed in a couple of seconds, by which time he'll have covered about 10 metres. As your coach said, if you can't stop them in the first few metres, you'll have trouble stopping them at all. (Don't take these numbers literally. Our simple model says that you'll be up to top speed in 'a few seconds' after 'a few metres'. Saying you'll be at top speed in 1.26 seconds after 5.83 metres is far too precise and is not appropriate, given the crudeness of our model. It is satisfying, though, that the physics describes reality reasonably well).

You can see the consequences of this piece of physics in every game. When on the attack, stagger the backs at a steep angle. This ensures that when the ball emerges from the scrum, ruck, or maul and is passed along the line, your backs have had time to accelerate and so receive the ball at close to top speed. The model also confirms why, on defence, the backs should line up flat along the offside line: you want to give your opponents the least distance to run with the ball before you tackle them.

In the 2005 game against Wales, England forgot this basic principle. Mike Ruddock, then Welsh coach, claimed that Wales had the best backs in the world — the challenge was to get the ball to them. Long kicks from England's backs, ones that failed to make touch, gave the Welsh backs the possession they craved. They could get up a full head of steam before crashing through the English defensive lines. Gavin Henson and Shane Williams are tough to stop even before they get to top speed. England forgot what they were taught in high school and suffered the consequences.

Our simple version of Hill's equation for sprinting gives plausible answers. His actual equation uses a resistance that is some constant multiple of the runner's speed, so that $R = kv$. When you run at top speed, call it u, we know that $R = F_{MAX}$, for there is no acceleration (otherwise it wouldn't be your top speed!) In other words, $F_{MAX} = ku$ or $k = F_{MAX} / u$. The equation Hill makes us solve is:

$$ma = F_{MAX} - kv = F_{MAX}(1 - \frac{v}{u})$$

This is simple enough that it can be solved exactly, without the need to resort to a computer[1]. The answer is:

$$v = u[1 - \exp(-F_{MAX}t / um)] = u[1 - \exp(-t / T)]$$

This is a clean, elegant equation. To make things better, it is easy to check. Studies have been done on American sprinter Carl Lewis, who won nine Olympic gold medals for running and long jump. Armed with a radar gun, it's straightforward to work out Carl's speed v any time after the start of the race. The radar gun and a chronometer let you draw a graph of v as a function of t. (See Figure 2.1) The biggest value of v that you get is u. Scientists plot the graph and then choose the value of $T = um / F_{MAX}$ that gives the best fit to Hill's equation. Since we have measured u and T, we can work out the value of the ratio m / F_{MAX}.

To check the model, note that the world record for the 100m is 9.69 seconds, so the average speed is 100/9.69 ~ 10.32 m/s, and we expect u to exceed this. Also, we predict that $T = um / F_{MAX} \sim 10m / mg \sim 1$ sec. At the World Championships in Rome, 1987, Carl ran the 100 metres in 9.86 seconds. A radar gun measured his top speed as 11.65 m/s and the best fit to the Hill model was when T equalled about 1.25 seconds. Because of the exponential in Hill's equation, after $2T$, about 2.5 seconds, Carl is running at about 86% of his top speed. His values of T and u give the precise ratio $F_{MAX} / m = 9.19$ N/kg, impressively close to our prediction of 9.81 N/kg.

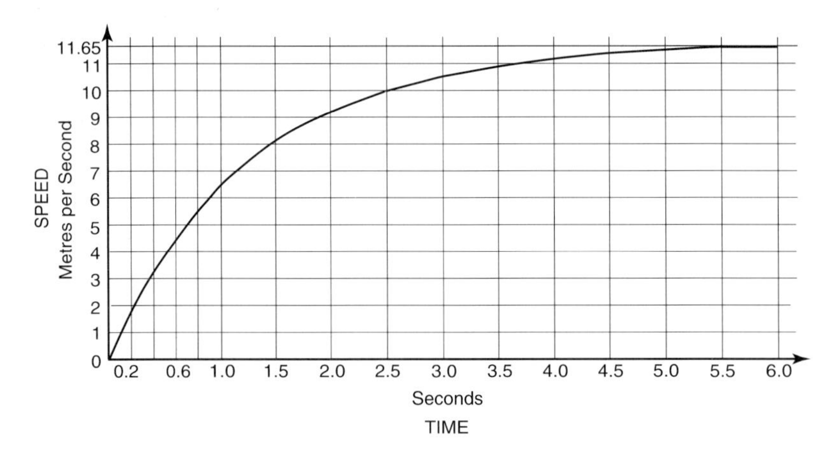

Figure 2.1: Sprinting speed as a function of time.

This precise mathematical model and the experiment confirm our intuition: that F_{MAX} is roughly equal to mg. This certainly should hold well on the track, where athletes have spikes and the track surface is engineered to help prevent loss of grip. Our simple model for JPR, which averaged the force and allowed us to use constant acceleration, had him reaching top speed in about 2 seconds; Carl is 86% of the way there in about 2.5 seconds. The moral of the story is that our simple model, where the runner exerts a constant force, works well. The high-powered mathematical model works better, as it should, but the main principles and features remain the same.

We can do more. Hill's equation for the speed can be solved to give the position of the runner. JPR has travelled a distance x in time t where:

$$x = \frac{F_{MAX}T^2}{m}\left[\frac{t}{T} + e^{-(t/T)} - 1\right]$$

This equation states the obvious: the faster you can run and the quicker you can get up to top speed, the more ground you can cover. To go as far as possible in the least possible time, you need F_{MAX}/m and u to be big. Find someone to teach you to run efficiently, for this will increase u. Building up those legs, arms, and abs will also help to improve u and, for that matter, F_{MAX}. (Many years ago, I saw the Bedford Blues training, which involved the backs sprinting up a short, steep slope while bearing hefty gentlemen of the scrum on their shoulders. It's a great way to build up those calf and quad muscles).

As a rough estimate, using Carl Lewis's values of T=1.25 and F_{MAX}/m=9.2, we expect that one second after catching the ball, a fast-paced back will have gone about 3.5 metres. After two seconds, though, our back has gone a total distance of 11.5 metres since receiving the ball. If you want to stop them scoring, hit them hard and early!

OF ACCELERATION AND SPEED

Speed is only one part of the game. Another important item for backs is acceleration. Your opponent's pace may be quick, but is his acceleration blistering? Can we tell who'll be quick and/or who can accelerate? A back-of-the-envelope calculation provides answers. The force that a muscle can exert is determined, so professors of biomechanics assure us, by the cross-sectional area of the muscle. Call this A. Now suppose the back's leg is of length L. The mass of the leg is proportional to leg volume, which is AL. Since acceleration is force divided by mass, the acceleration that the leg can provide is $a = F/M \propto (A/AL) \propto 1/L$. If we say that leg length is about half your height, then the smaller the back, the more rapid his acceleration is likely to be. So, Scottish fly half Phil Godman (at 1.78 m) can accelerate 6% faster than his teammate, fullback Rory Lamont (1.88m), and 11% faster than lock Nathan Hines (1.98m).

While we pause to ponder the inside leg measurements of some of rugby's greatest runners, let us think for a moment about the simple pendulum. We want to know how fast we can run, how to maximise u in the Hill equation. To do so, we can contemplate the problem while pacing up and down, waiting for inspiration. While pacing, plant your left foot on the ground and swing your right foot forward. Then plant your right foot firmly on terra firma and swing the left foot forward. With sudden inspiration from the swing of the leg, we model running as though the leg is a simple pendulum. And, to paraphrase a Clint Eastwood movie, there are two types of people in the world: those who know that a simple pendulum of length L takes a time T to oscillate, where $T \propto \sqrt{L}$, and those who don't. You are in the former category[2]. The length of the step, your stride length S, is proportional to L. So, your running speed goes like $S/T \propto (L/\sqrt{L}) \propto \sqrt{L}$. And, once again, if your leg length is proportional to your height, we predict that tall players run faster. Yet again we can check the model: The 100 and 200 metre world records are currently owned by Jamaica's Usain Bolt, who is 1.96m tall (about 6' 5') which most people would agree fits the 'tall' category.

These estimates suggest what needs to happen. Scrum halves and fly halves need to accelerate rapidly into gaps (after all, fly half is a contraction of 'flying half' so we'd expect a good deal of acceleration from our number ten). The scrum half has the job of blasting his way in for a try from a five metre scrum, while the number 10 has to sprint into space to clear the ball downfield. But if the time is ripe for sending the ball down the line, then you want it to go to people who are extremely fast and who have had enough time to get up to top speed. So, among the backs, we predict that numbers 9 and 10 will be relatively short, while the centres and full back will be tall. Vive La France!: Scrum half Jean-Baptiste Élissalde is 1.72 m while their centre Yannick Jauzion is 1.93m. For wingers, all bets are off. You can pick either Jonah Lomu (1.96m) or Shane Williams (1.73 m), both are reasonably effective!

GIVING THEM THE SLIP: FRICTION

Speed and acceleration are fine, but if you've ever played on a slippery surface, you'll know all about traction. Basically, put too much effort into pushing off the ground and your foot will slip. The reason, oddly enough, is the molecules of which the bottom layer of your boot and the top layer of turf are made. As you plonk down your size 10 on the hallowed turf of Twickenham, molecules of mud and grass form temporary chemical bonds with those of the rubber sole of your footware. To move your foot, you have to break those weak little bonds. To break those bonds requires a force, and the force exerted by those weak molecular bonds is known as friction.

Friction is baffling, for it does not depend on surface area one bit. Phil Vickery's clod-hopping size 12 generates no greater frictional force than that of twinkle-toed Argentinian fly half Juan Martin Hernandez. The friction force *does* depend, however, on the two surfaces involved. In a high-school physics class, rough blocks are forever laying on flat surfaces. The blocks have mass M and the coefficient of friction between the block and the surface is f. You will not escape the classroom until you know that the force you need to overcome friction and get the block to move is fMg and that f is at most 1.

This should generate a health warning to rugby players everywhere: If, when trying to run, you exert a force greater than fMg, your foot will slip, for you will lose traction. For those who love definitions, $1-f$ goes by the name of the slipperiness, for obvious reasons.

If rugby were played on artificial turf (think of the carpet burns) then we would have f almost equal to 1. We'd run faster, but fall harder, and may end up needing the armour that is a requirement in American football. Luckily, we play on a softer, friendlier surface.

Fortunately, the interaction of rubber and grass is closely studied. The coefficient of friction between them is about 0.35. So, the maximum force you can exert without slipping is around $mg/3$. In principle, then, the maximum value of F_{MAX} / m is $g/3$. On a wet field, the coefficient of friction plummets to about 0.2, in which case $F_{MAX} / m = g/5$. We know these coefficients well because the police use them in accident investigations. If your car goes off of the road onto the grass, they can measure the length of your skid marks. This lets them know how far you travelled once you'd hit the brakes. Knowing this distance, and the coefficient of friction between wheel and grass, the police can work out the minimum speed you were going prior to the accident and know full well if you were breaking the speed limit or not.

Your top speed, we hope, is unaffected by the conditions of the turf. What will change, though, is the time it takes to there. If you can do so in about 2 seconds on perfect, nonslipping surfaces, it'll take you 6 seconds at Twickenham and 10 seconds at Murrayfield. You'll be far easier to catch. To avoid ruining your prospects for an international career, scientists have made a breakthrough: the stud. With studs, you put all of your weight not on the whole area of your number 12 boot but on a set of, say, 6 studs. If your foot is 12 inches long and 4 inches wide, your feet have a total area of about 100 square inches or about 625 square centimetres. You are required by the rules of the game, Law 4 for those of you who doubt, to have studs of minimum diameter 13 mm. The surface area with studs is a mere 16 square centimetres, a drop by a factor of about 40. The net result: you puncture the surface with ease. But once the studs are in the ground, they act as little anchors. You can exert much more forward push without slipping, so you can increase F_{MAX} / m. So, to run swiftly, invest in a few sets of studs of differing length, using longer ones for greater purchase when the ground gets sticky, or going for a ridged sole when playing sevens. That way, you increase the traction and maximise the force you can exert on the ground without slipping, so you increase F_{MAX} / m thereby running faster and avoiding injury. Be warned: scientists have shown that boots where the studs are arranged around the edge of the boot with none in the middle (typically a football boot rather than a rugby boot) have great association with knee injuries, in particular the tearing of the anterior cruciate ligament (the ACL). Check the soles the next time you buy boots.

Why do the rules specify a minimum stud width? Because smaller stud sizes would increase the pressure you exert even more, and any contact with the body of an opponent could result in devastating injuries. For example, Brent 'Who Me, Ref?' Cockbain, has a mass of about 122kg. When this is exerted over 12 studs of the minimum 13 mm diameter, it amounts to a pressure of 750 kPascals or, in nonmetric terms, about 120 pounds/square inch. For comparison, the tires on a car are about 30 pounds/square inch. Think of the damage Mr. Cockbain could inflict accidentally should he step on your prostrate body at a ruck.

Beyond boots

With our boots bought and our studs selected, we can return to running. We can embellish Hill's simple sprint model. First, we've assumed that the force is proportional to your mass. But what if it depends on your 'useful' mass – the bones, tendons, sinews, and muscles? This would change our model slightly. All we'd need to do is to replace F_{MAX} by $F_{MAX}[1-(b/m)]$. As before, m is your mass but b is your body fat (blubber!) that you carry. The higher your body fat percentage, the less force you can put into running. A fit rugby player might have about 6% body fat. At 25%, labels such as 'obese' are applied and it will take you about 25% longer to get up to top speed. The moral is simple: lose the extra pounds.

Resistance at low speeds is a simple multiple of the speed, so Hill tells us. Once you're cruising along, though, air drag enters into the picture. This depends on the square of the speed, so if you double your speed, you quadruple the drag. It also depends on the area of the player. That's how a parachute works: the large area causes a large drag force, so that the top speed, the terminal velocity, is a gentle 10 m/s. This is about the same speed you'd reach if you jumped, without a parachute, from a height of 5 metres.

People who design cars and airplanes suspend plates in a wind tunnel and blow air past them at high speed. The drag force on the plate depends on its width as well as the angle at which that plate is suspended. You might not want to believe that your backs and three quarters resemble solid rectangular plates, but in a certain sense they do. The wind tunnel teaches us that if all of your backs have the same muscles and exert the same force, the solidly constructed backs who lean forward as they run will do better than the gangly youths who run upright and flap their arms about them as they sprint down the pitch. You also want to wear kit that fits. Not only does a skin-tight jersey give your opponents less chance of grabbing your clothing as you pass, it also reduces your cross sectional area and so lowers air drag.

Sultans of the Sidestep

Phil Bennett was one of the world's greatest fly halves. He played club rugby for Llanelli — most memorably on 'the day the pubs ran dry' when the Scarlets beat the visiting All Blacks 9-3 at Stradey. Phil was great with his feet, whether kicking the ball or dazzling opponents with his fleet-footed change of directions. He played a starring role in the greatest try ever scored (you can see it on youtube), when Sid Going and his 1973 All Black's took on the Barbarians at the Arms Park. Gareth Edwards dived over for the score, but it's the fantastic footwork of Phil that is burned in the memory. The Baa Baa's ran out winners that day 23-11, but 'The Try', a mere four minutes into the game, is what you remember, not the final score.

Acceleration, lest we forget, can mean running in a straight line more quickly, or running at the same speed but changing direction. Our simple model of acceleration suggests that backs at the shorter end of the spectrum can accelerate the most swiftly, so they should be best at the side step, assuming their boots give them good traction. Phil Bennett is 1.70m (5' 6') tall while Shane Williams is 1.73m (5'7'), suggesting our predictions are pretty good.

There's a branch of physics that can model the behavior of twinkle-toed players who leave everyone in their wake. It bears the name of a Scottish biologist, Robert Brown, but its main impetus came from a physicist by the name of Albert Einstein. It's not widely known, but Einstein's most frequently referenced work is not on special or general relativity, but on Brownian motion.

Back in the early 1800s, Brown noticed that pollen grains, if you look at them under a microscope, have a strange, zigzag motion – not unlike Phil Bennett on the rugby pitch. Einstein eventually explained this motion in terms of the laws of physics. This motion, thanks to Albert's explanation, became one of the key pieces of evidence that atoms really do exist.

To return to Bennett, Shane Williams, and other jinking, ducking, and weaving players, we need only toss a coin. Suppose Phil gets the ball on his own try line – as he did with the Baa-Baa's on that famous day. Bizarrely, he has a coin in his pocket. The All Black's line is directly North. Phil runs straight, as a good back should, until one of the Kiwis get close. Phil whips out the coin and tosses it. If it's heads, he takes a step northeast; if tails, northwest. (Mind you, coin tossing in rugby is not easy; the story goes that Princess Ann's son Peter Phillips, when captain of the Gordonstoun rugby team, was asked by a ref who was probably petrified by protocol whether he wanted 'grandmother or tails.')

We, however, want Bennett to toss the coin and see what effect it has on his ducking and weaving down the pitch. After all, would he have been better off following the example of Crashball Kent and simply crunch into his opponent? If we have to defend against Phil or, these days, against Shane, can we get any idea of where to position ourselves?

Suppose Phil catches the ball right on his own tryline, right between the posts. If he steps on the accelerator, then he'll cover the 100 metres to the opposite tryline in about 67 strides, for each stride is about 1.5 metres, and this will take him about 10 seconds. To make life easier, suppose one of the opponents has been carded for some type of naughtiness (Phil played before the days of carding, but you get the picture). The enemy flock towards him and so, to avoid capture, Phil has to side step 14 times. Before heading North East or North West, he tosses a coin. (See Figure 2.2) If the coin is fair, heads will come down seven times and tails for the other seven. If that's the case, then on average he will plonk the ball down slap bang in the middle of the posts, for the easiest conversion possible.

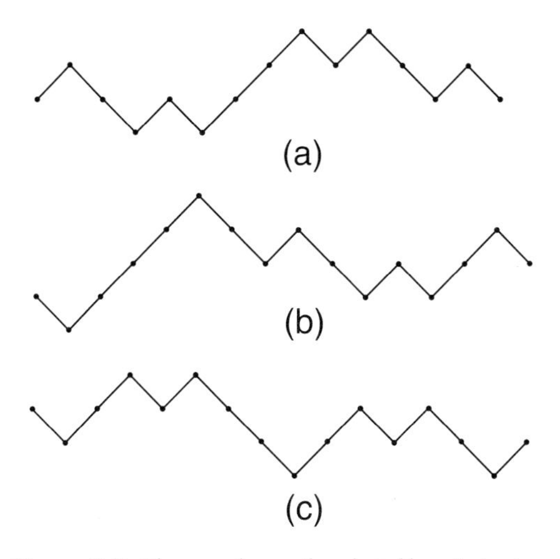

Figure 2.2: Three random paths selected by coin tossing. Each resemble the jinking motion of a master side stepper.

When Phil sidesteps, his 1.5m stride is no longer purely North. In fact, the stride takes him only 1 metre closer to his opponents tryline. He sidesteps 14 times in all, which carries him about 14 metres towards the line. He has another 86 metres to go, which is about 57 strides. When Phil runs straight, he covers the pitch in 67 strides; when he ducks and weaves, it will take him 71 strides. As Phil can run at 10 m/s and a stride is 1.5 metres, his speed is 6.7 strides per second. So, in a flat out sprint he goes coast-to-coast in ten seconds. When sidestepping 14 times to avoid the outstretched arms of the enemy, it will take him 71/6.7, or about 10.6 seconds. If we add in a quarter-second hesitation (the time he takes mentally to toss a coin and decide which way to go) as part of the side stepping process, he'll cross over for the score in about 14 seconds.

The jink that takes him 1 metre down the pitch also takes him 1 metre across the pitch. Our model of heading North East or North West is not a bad one, for the width of an outstretched arm is about 1 metre. If an All Black sticks out an arm to grab him, Phil will evade the tackle.

The maximum width of the pitch is 69 m. If Phil heads Northwest with every toss of the coin, then he'll be about 20.5 m from the centre of the pitch. (The worry, of course, is that things change. In 2004, the Australians complained bitterly after having played at the miraculously shrinking Murrayfield. The Wallabies trained on the Scottish pitch which was set out for the usual 69 m – the maximum width according to the rules. On game day, though, it was a cramped 65 m wide. On such a pitch, Phil's 14 skips North East would take him to about 18 metres of the touchline. He still has ample room, but any All Blacks who chase him down now have an easier target.)

As he goes downfield, Phil makes fourteen decisions. That means he has a phenomenal 16,384 (= 2^{14}) ways to get from one end to the other – at least in terms of deciding which direction to sidestep. That's the advantage of sidestepping: it may take Phil a few seconds longer to duck and weave down the pitch, but his path is massively unpredictable. If you know where he's headed, you can try to cut him off. If you don't, then life is much harder. But even though he's got 16,384 possibilities, physics lets us hunt him down.

Phil starts halfway across the pitch. After N sidesteps, he's a distance R to one side. The square of this distance is useful, for we no longer have to worry whether he's to the right of left of the pitch; we care only how far he is from the middle. The average of this quantity, the so-called mean square distance, is written $< R^2 >$, since the time honoured tradition is that <…> means the average of whatever lies between the angled brackets. The theory of Brownian motion tells us[3] that in our example, $< R^2 >= N$. This helps you to defend. Phil can, in principle, duck to the left 14 times in a row, taking him 14 metres left of centre. Or he could go right 14 times and end up 14 metres to the right of centre. This leaves you about 28 metres of turf to protect, rather a daunting task. But if our model holds, and that's a big if, then Phil's most likely to end up within about $\sqrt{14}$ or 3.74 metres of centre. Running and weaving at random will not take him too far from the middle of the pitch. If the defence is both fit and desperate, so that each of them tackle him twice, then N=28 and $\sqrt{< R^2 >} \sim 5.3$. So, the more tacklers he faces, the farther he is likely to drift from the middle of the pitch.

This simple model of a darting run is pleasing. It gives a reasonable time for a back, given the ball in the middle of the pitch, to score a try. The erratic path of this coin-tossing process mimics that of a back running into heavily guarded territory. Last, it predicts that Phil inexorably drifts towards the touchline – and we all know how difficult it is to score under the posts. To use a phrase from the lexicon of physics, I need to add some 'weasel words.' Our math holds true only on average, when we average over tens of thousands of decisions. As Phil has at most fifteen decisions to make, we've taken some liberties with the mathematics. So be it – our model seems pretty good. Also, it assumes that Phil chooses at random, when it fact he's more likely to dive to the left if his opponent is slightly to the right, or if Phil has a heavily favoured foot on which to sidestep.

There is another true-to-life part of this physics. Just because Phil has sidestepped to the left four times in a row does not mean he simply *has* to go right next time. In a coin tossing process, long runs of heads – or tails – are all part of the mathematics. Anyone who thinks that the more you head to the left the more overdue you are to go to the right is swallowing the 'gambler's fallacy.' Many a gambler has lost large amounts of money by believing that a long run of reds in roulette means that, surely, the next time black must come up. The mathematical theory of runs, how many runs of N heads in a row will crop up if you toss a coin M times, has applications in diverse areas such as testing whether slot machines are biased, radioactive decay is truly a random process, and whether items produced on a production line are acceptable or not[4]. In rugby, it tells us to be wary of guessing which way Phil will run. Just because he went to the left four times straight does not mean that he has to go right next time.

All you need to create a virtual half back is a random number generator – and there's one in Microsoft Excel [the RAND(...) function, for doubting Thomases]. This is a good thing. In the Second World War, a prisoner of war, statistician John Edmund Kerrich, had time on his hands. A Briton who had emigrated to South Africa, Kerrich was imprisoned by the Nazi's while on a visit to his wife's family in Copenhagen. To while away the hours, he began tossing a coin and recorded the results. After 10,000 tosses, he obtained 5,067 heads and 4, 933 tails. Most humans could not face the tedium of tossing a coin ad nauseam. Computers don't mind – they excel at boring tasks. To make an artificial Shane Williams, get the computer to spit out random numbers between zero and 1. If this number is less than 0.333334, make Shane go Northeast. If it is more than 0.666667, he moves Northwest. Otherwise, he goes straight ahead. (See Figure 2.3)

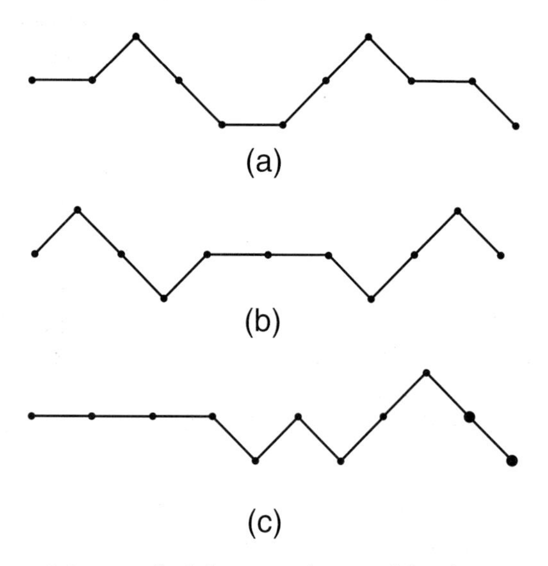

Figure 2.3: An artificial Shane. Here he can go left, right, or straight.

Those who feel so inclined can construct a more complex system. Allow the runner to proceed at speed v but forget Northwest and Northeast. Let the angle at which he darts be random at every instant[5]. The basic reasoning behind this model is when a fast but small back batters into heavy, motionless forwards he may bounce off of them like a ball against the barriers in a pinball machine. He is not tackled, instead he merely bounces off of them without losing speed, undergoing only a change in direction. The key feature of this model is that, soon after you start, your motion is similar to straight-line running; after a while, however, you look just like the Brownian motion model.

This 'constant speed, ever changing angle' model bears some relation to how physicists describe polymers (molecules that can be thought of as long chains of links) and proteins (more closely akin to a rope thrown on the ground at random). For a polymer, the model has a special number that marks how easy it is for the long chain to snake back upon itself. In rugby, it might be a measure of willpower – how much the fleet-footed back truly wants to go straight ahead!

It is sad, but true, that Einstein never turned his attention to rugby. He looked at coin-tossing behavior to understand how molecules of one type would spread out when placed in a liquid or gas that contained different types of molecule. He modelled the hopping of molecules back and forth in much the same way as we theoretically sculpted equations to describe Phil Bennett's motion[6]. Our equation for the root mean square distance, which increases as the square root of the number of steps and thus as the square root of the time that has passed, is a profound result. It lies at the heart of diffusion processes. Ink that spreads out in water, for example, is classic diffusion behavior. The amount by which the ink spreads goes as $\sqrt{t}$. If you pour boiling water into a cup, let it settle, and then put a tea bag in it, the brown smudge spreads out according to the $\sqrt{t}$ law. Mind you, if you stir the tea, all bets are off: that causes *turbulent* diffusion to take place, which is much more rapid.

1905 was Einstein's year of miracles. That year, he published a paper that gained him the Nobel Prize, two papers that founded the special theory of relativity, and he also published his most cited article ever: his paper on

Brownian motion. Einstein's theory described not only how a liquid diffuses with time, but also how that diffusion depends on temperature. It has been enormously successful, and is put to good use not only in physics, but also chemistry and chemical engineering. In 1907, two years after his pioneering work on Brownian motion, Einstein re-wrote it in a less mathematical form. This was for the benefit of chemists, who might not have the same mathematics training as physicists. It paid dividends. To check Einstein's theory, French scientist Jean Perrin devised some extremely time-consuming, detail-intensive experiments. For these observations, which confirmed Einstein's theory in every detail, Perrin won the Nobel Prize in physics in 1926, five years after Einstein. It was awarded for 'his work on the discontinuous structure of matter, and especially for his discovery of sedimentation equilibrium.'

RUNNING WITH THE HERD

Aussie David Campese and All Black Jonah Lomu have helped reshape rugby. Nick Farr-Jones said of Campese 'he's the sort of player whose brain doesn't always know where his legs are carrying him." Gavin Hastings, with famous Scottish bluntness and far less poetry than Farr-Jones, described Lomu with the words 'There's no doubt about it, he's a big &*%#'.

In the modern game of rugby, though, it's not just the backs like Campese and Lomu who need to move swiftly. Packs have become far more mobile than ever before. Props now need to get across the pitch in a hurry, too. In the 70s and 80s, packs were huge and lumbering. Now they are huge and fast. It's not rare these days to see a prop run, and catch, a three quarter. Duncan Jones, one of the Osprey's Hair Bear props is a case in point. Mobility has become the name of the game.

Physics also has a way to describe the sweating herd of humanity that forms the pack. It's a different model from the one that describes the sprinting of the backs. The hero of the model is Pheidippides. According to legend, after the Battle of Marathon in 490BC, he ran from Marathon to Athens to announce the great Greek victory over the Persians. He arrived, said 'We are victorious,' then died. Ancient sources say no such thing, but it remains a popular story.

The legendary Pheidippides is a role model for forwards – running, grunting, pushing, shoving, and having only enough energy when the whistle blows to shout 'We are victorious' before collapsing. All coaches want players who, no matter how daunting the odds, no matter how much effort they've already spent, won't stop running until the final whistle. The model for how pack members run requires them to collapse the second the game ends. In other words, by the end of the game they are drained of their entire energy.

But what is this energy? To construct another back-of-the-envelope model, strip down energy to its bare minimum. Our pack motors around the park, each member having a mass m and speed v. The kinetic energy of every player, then, is $\frac{1}{2}mv^2$. This drains energy from Dunc, for in order for this bruiser to cruise at this speed, he has to overcome resistance with every single footfall. The rate at which his kinetic energy drains is Fv, where F is the force he needs to exert to run at speed v and which – unlike sprinting – is not F_{MAX}.[7]

Luckily, Mr. Jones's body produces more energy, chemical energy, which he does by burning fat at some rate S. For a physicist, power P is the rate at which you produce energy E. A light bulb, for example, is labelled as being so many Watts. A 100-Watt bulb puts out 100 Joules of energy every second. Jones and the other Ospreys expend a power P by burning energy at a rate S and overcoming resistance to motion. To keep things in check, we have:

$$P = Fv - S$$

Our barebones version of the Hill-Keller endurance model is summed up in the simple phrase: the power is constant. If you start off with an energy E and after time T you have no energy left, then you emit $P = E/T$.

We can do more. The force F that you use has to overcome a resistance to motion kv. This causes an acceleration, so that $ma = F - kv$. We want to work out what a reasonable running speed is for one of the forwards. Are they lolling around doing nothing much of anything, or are they really trying their hardest? What should our expectations be?

Suppose the Saracen forwards run at constant speed. They've all played in horrible games when the backs of both teams thump the ball with their boot, so the pack spends the afternoon running 60 metres for the lineout at their end of the pitch, then running back the whole sixty metres for the lineout at the other end. And so it goes. If their speed is constant, then the acceleration is zero and the force F that they use to run equals kv, the resistance to motion. What is the power that that Andy Saull and each member of the Saracens put out?

To answer, think of the Marathon – the race, not the battle – as an example. It is an endurance event and, according to our model, is run by someone who exerts a constant power. Suppose this power generated is the same for the marathon as it is for a Rugby World Cup game. If both Saull the Saracen and the marathon runner generate power from their muscles in the same amount S, and if the resistance to motion k is some multiple of the runner's mass, then we can check our theory, which reduces to saying that the kinetic energy of rugby player and marathon runner are the same. We have:

$$\frac{m_A}{m_R} = \left(\frac{v_R}{v_A}\right)^2$$

where R stands for runner and A for Andy. A top marathon runner covers the 26 miles 385 yards in just over two hours. (The odd distance is a result of the London Olympics of 1908, where the distance was increased from 25 miles so that Princess Mary could see the race begin at Windsor Castle from the Nursery window.) The distance is roughly 42,000 metres, so that a modern world-class runner has a speed of some 5.8 m/s.

The current world record holder in the Marathon is Ethiopia's Haile Gebrselassie, who tips the scale at 56kg. If his power output is the same as John 'Bull' Hayes, the Munster, Ireland, and Lions prop who weighs in at a muscular 125kg, then Bull can cover the pitch with an average speed $v = 5.8\sqrt{56/125}$, so that $v = 3.9$ m/s, a tad shy of 9 mph. This is about three times as fast as you can walk, so our model seems reasonably good. A more lightweight forward, such as Bull's fellow countryman and Northampton Saints flanker Neil Best – a mere 105kg – could manage 4.2 m/s, getting on for 10 miles per hour. This is not a bad estimate. It suggests that the leisurely pace of the pack carries them one mile in six minutes, about half as long again as Sir Roger Bannister's famous run at Oxford in 1954. Given that the pack have other things to do – such as push, tackle, and what not – 3.9 m/s is pretty good.

While the teams run around the pitch playing superb rugby, folks in the stand do precious little of anything. If we're normal human beings, we should only consume about 2,000 calories per day, lest we get fat. Strangely, what are called calories are, in fact, kilocalories, and one kilocalorie is about 4.18 kilojoules. Doctors say to restrict our energy intake to about 9.6 million Joules per day. We burn these Joules up over 24 hours. If we don't move, our speed is 0, and the rate at which we burn energy must be S. So, to estimate S, we figure it must be 2,000 kilocals/day or, in metric units, about 100 Watts. The power put out by our rugby players is $P = kv^2 - S$. Our minimum estimate for S is going to be 100W, for your body has to produce more energy to play rugby than it does to sit watching rugby on television. We know that the resistance coefficient $k = m/T$, where m is the mass and T is the time it takes to get up to speed, which is about a second. If Bull Hayes lumbers up to 4 m/s in about 2 seconds, then his power output during the game is some 900 Watts, just shy of the kilowatt electric fire you might huddle around on a winter evening to keep warm. Some might prefer, therefore, to huddle around Mr. Hayes though I, for one, would not attempt it.

This wattage is a bit high. The most elite athletes in the world can produce a burst of 1 kilowatt for a few seconds at best. Still, our incredibly simple model of rugby power suggests players produce the high-hundreds in terms of wattage, which is a pretty good estimate.

During an 80 minute game, a power output of 900 Watts produces about 4.3 million Joules or 4,300 kiloJoules. That's about half the energy a couch potato burns up in a single day. Rugby players, it seems, need to keep their calorie intake high. One way to replenish the 4,300 kJ burned up during the game would be to consume a Big Mac, whose nutrition information pegs it at about 2,000kJ. Those who don't like the taste might then choose to wash it down with four cans of beer (585 kJ each) for a grand total of 4,325kJ.

More seriously, pure fat has an energy content of about 38 kJ per gram. During the game, then, you'll burn off about 113 grams, or about a quarter pound of pure fat. Your actual weight loss would exceed that, for a great deal of water is lost by sweating.

So, the Hill-Keller model of endurance tells you to stock up before the game on water and energy supply. Rather than beer and Big Macs after the game, you can get reserves of 4000 kJ by eating about 280 grams of protein or carbohydrates the night before. This is otherwise known as carboloading.

As with the model for sprinting, our crude description does a good job explaining what needs to be explained. The power output of a player is a tad high, its true, but the kilocalorie/kilojoule consumption and the running speed look about right. The full-blown model for endurance events, the Hill-Keller model, is an extremely advanced equation from the realm of mathematics that requires some high-powered theoretical methods to solve it. Rather than spend the effort to solve such a daunting equation[8], our result is good enough.

PASSING

The application process at Oxford University's St. Edmund Hall is, according to legend, simple. As you walk through the doorway, a rugby ball is thrown towards you. Catch it and you'll be accepted; pass it, and you'll get a scholarship. Students at Teddy Hall, as it is known, are renown for their rugby prowess. But no matter where you study, how strong, how fleet of foot, or how dazzling your sidestep may be, the time always comes when the best thing to do is to pass the ball.

THINKING BACK, PASSING FORWARD

The first thing students of the game are taught is that in rugby the ball must go backwards. It is vexing when your team is penalised because the backs passed the ball forward – especially if you're having a rough ride in the scrums. As a fan, it is annoying when the referee constantly blows the whistle for forward passes – are the teams bad or is the ref awful? Physics shows us the way. The problem with passing is that it is truly relative.

Suppose a player has the ball and is running with a velocity $\vec{v}$. This is a vector, something with both size and direction. It is the velocity of the player, say former Saracen and French centre Thomas Castaignede, *relative to the ground*. He runs straight down the pitch, so his speed is v towards the Irish line. Thomas launches a superb pass, at an angle A *as he sees it* at a velocity $\vec{u}$. According to him, the ball has a speed *across* the pitch of $u \sin A$. So far, so good. Thomas errs because he thinks the ball has a speed, headed back towards his own line, equal to $u \cos A$. Monsieur Castaignede believes it's a fair pass; that's what he'd write in his rugby column for *The Guardian*. But those of us in the stands beg to differ. We are perfectly stationary, so we see something different. On top of the $u \cos A$ towards the French line that Thomas perceives, we add on Thomas's own speed. In other words, the crowd sees the ball travel at $v - u \cos A$ towards the *Irish* line. Unless Thomas can make $u \cos A$ bigger than v, the Croke Park crowd will roar that the pass is forward. Castaignede has to throw the ball at an angle A so that $\cos A > v / u$. (See Figure 2.4)

This gives the first rule of passing: throw hard. If you do not make the ball travel quicker than you run down the pitch, then *v/u* will *always* be greater than 1 and – as cos A is always less than 1 – there is absolutely no way that you can throw it so that a stationary ref won't blow the whistle for a forward pass. This may seem crazy, but suppose Castaignede runs at 10 m/s and passes the ball directly behind at 9 m/s. To Thomas, the ball is obviously going backwards, at 9 m/s relative to him. To a fan in the stands, the ball is going *forwards* at 10 – 9 which is 1 m/s. The crowd howls, the Greens scream, for the French have scored a game-winning try from a pass that was clearly forward – to them. If the referee is unfit and is stationary, he will agree with the Dubliners and blow his whistle. On the other hand, if the referee is fit enough to move at the same speed as Castagneide, his arm goes aloft and he awards Les Bleus the five points.

One solution for referees: look at the lines marked on the pitch. If the pass happens close to the 22, 10, half way, or goal line, you should be able to use the line to see whether, relative to the ground, the ball went forward, backward, or sideways.

But if you've just scored, only to have the official disallow it for a forward pass, don't get mad. Just think about Jonathan Davies view of rugby, 'I think you enjoy the game more if you don't know the rules. Anyway — you're on the same wavelength as the referees.' Mind you, Scottish fans still burn from the forward pass that let France score a game-winning try in their 2009 Six Nations match.

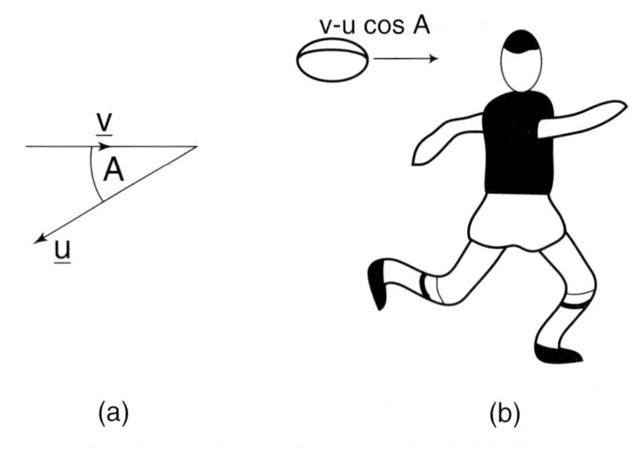

Figure 2.4: In (a), the player thinks he's passed the ball backwards at speed u and angle A. In (b) the crowd sees the ball moving forward at v-u co sA.

STIFF ARMS, DIVE PASSING AND LAMB CHOP

Aspiring internationals do not wait until the ball floats gently towards them on the breeze; they reach out to catch it. Then, a swift movement of the arms sends the ball hurtling from fly half, to inside centre, to outside centre, to Shane, who plops the ball down for the full five points.

A crude, yet satisfying, model of a pass would be this. Extended your arms horizontally to catch the ball. Once caught, the arm-ball combination rotates freely under gravity, just like a pendulum. (See Figure 2.5) Once the pendulum reaches some angle A, let go of the ball. How far can you throw it?

Letting your arms fall limply under gravity is no basis for playing real rugby, but the model is a good place to start. As you extend your arms to catch the ball, both the ball and your arms are at height 0 – roughly equal to the height of your shoulders above the ground — and move at speed zero. Let them drop under gravity. You convert gravitational energy into kinetic energy. Once your arms make an angle A with the vertical, let go of the ball.

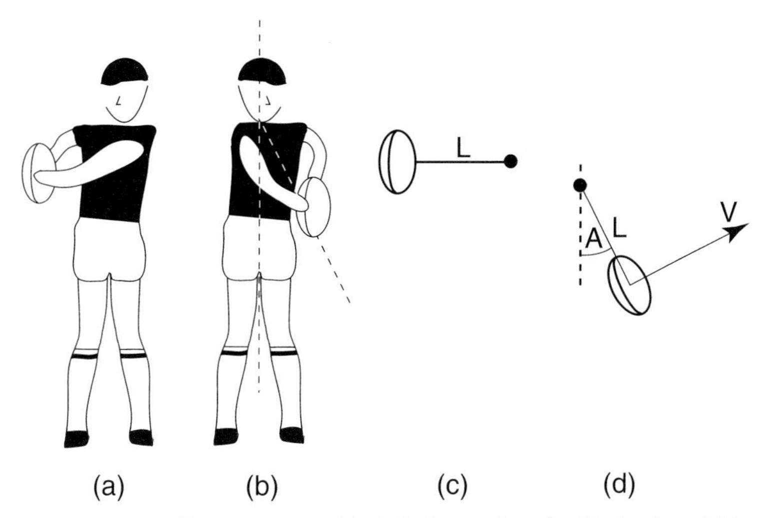

Figure 2.5: Stiff arm passing. In (a), the ball is caught at shoulder height and (b) it is passed. In (c) and (d), passing is shown as a simple pendulum.

The instant you let go, your arms and the ball are rotating with a speed v. If your arms are of mass M and the ball of mass m, then the kinetic energy is $\left(\frac{M}{6}+\frac{m}{2}\right)v^2$. The potential energy, on the other hand, is $\left(\frac{M}{2}+m\right)gL\cos A$. Here, L is the length of your arm.

As energy is conserved, these two quantities, when added together, must be zero. This means that the launch speed v is given by:

$$v^2 = \frac{\left(\frac{M}{2}+m\right)gL\cos A}{\left(\frac{M}{6}+\frac{m}{2}\right)} = \frac{3(1+2x)}{(1+3x)}gL\cos A$$

where, to save ink, we've used the shorthand $x = m/M$. For any given launch angle, the fastest launch speed – and therefore the longest pass – is hurled by those people with the beefiest arms. If your arms are infinitely heavier than the ball, $v^2 = 3gL\cos A$. If your arms are puny, then $v^2 = 2gL\cos A$. So, work in the weight room to build beefy biceps. Studies — and trust me, you don't want to know how — suggest the average mass of a human's right arm is 4kg (the left arm is slightly lighter). In that case, $x = 0.1$, for the ball is about 425 grams. If your arms are about one metre long, then the launch speed is, at best, 5.2 m/s.

We can make more progress. In days gone by, the English and the French had different ideas about projectiles. The French believe that a cannonball, for example, would curve gently through the air, but they did not know what this curve was. The English believed that it flew horizontally until it 'ran out of air', at which point it plummeted. The English view was rather like the Roadrunner cartoons, where the coyote happily runs off of the cliff and continues to run horizontally. Then, when he realises the cliff is no longer there, he shrugs his shoulder and accelerates at 9.8 m/s^2 towards the canyon floor, kicking up a cloud of dust upon impact.

Isaac Newton, thanks to his laws of motion, found the curve that the French could not – one that slightly resembles the cartoon. The curve carved out by the Gilbert, or any projectile for that matter, is a parabola. To describe the rugby ball's motion through the air, if it is launched at an angle A with an initial speed v, you need two facts. First, the initial horizontal speed $u \cos A$ is undiminished as the ball goes on its merry way towards the outside half. The vertical launch speed is $u \sin A$, and motion in this direction is governed by the force of gravity.

When you throw the ball at an angle A, it will go up and – as what goes up must come down – it returns to the same height from which it was thrown (assuming passer and catcher are the same height). In so doing, it travels a horizontal distance R, the range, where:

$$R = \frac{v^2 \sin 2A}{g}$$

If the speed did not depend on angle, then the maximum range – the longest pass – would be when A is 45 degrees. We know, though, that the launch speed of our stiff-armed pass *does* depend on angle. For our model, the range is:

$$R = \frac{3(1+2x)L \cos A \sin 2A}{(1+3x)}$$

Another one of Newton's discoveries, calculus, let's us calculate the optimal launch angle, the one that maximises the length of the scrum half's shovel pass. The scrum half should let go of the ball when $\tan A = 1/\sqrt{2}$, giving a best launch angle of about 35 degrees to the vertical[9]. The longest pass will travel:

$$R = \frac{4(1+2x)L}{\sqrt{3}(1+3x)}$$

so that the best we can hope for is a pass of length $(4L/\sqrt{3})$ which is between about 2 and 3 metres. So, stiffly swinging your arms means you'll pass the ball a feeble distance and ruin your chances to play at the international level. The scrum half can pass farther if he's taller than the fly half, but it's not going to make the ball travel that much farther.

England has had, over the years, a number of great scrum halves. The list does not include Mike 'Lamb Chop' Lampkowski, who held that much-coveted position for four matches back in 1976. Lampkowski had trouble spin-passing the ball. Instead, he used to dive pass. (See Figure 2.6) As far as physics is concerned, there's not too much difference between the 'dead arm' pass and Lamb Chop's specialty.

Figure 2.6: The dive pass: The body speed of the player adds to the launch speed of the ball.

When you dive, your whole body, which is holding on to the ball, has a certain speed. If you hold the ball close to your stomach and then, as you dive horizontally, whip your arms down through the vertical position and out the other side, you can improve on the distance that the ball can be thrown. The ball has not only the stiff-arm speed but also your diving speed as well. If you can throw yourself your own body length - about 2 metres – your dive speed is about 6 m/s. Add this to the stiff-arm speed of about 5 m/s, and your dive pass can go about 7 metres. This doubles or trebles the range, but you lose a lot by being sprawled on the deck. The time it takes you to get up and running deprives your side of a try-scoring opportunity, or a missed try-saving opportunity. There's a reason England never won the Grand Slam with Lamb Chop at the base of the eight.

MEDIEVAL AND RADIOACTIVE PASSING

To add some oomph, there are ways to improve on this passing technique that could have been known by those who fought medieval siege warfare. Early on, rocks or the bodies of plague victims were hurled over castle battlements by a simple catapult device. A large counterweight was put on one end of a bar. Rocks were put on the other end, and the rock end was held down. Once you let it go, the counterweight plummeted downwards, launching the rocks upwards and onwards. It wasn't long, though, before the trebuchet came along. It had two major improvements. The counterweight was suspended below the bar, which added a quick last-minute flick to the rocks. Also, to extend the length of the bar, a sling was added. Rugby is somewhat like warfare. If we learn from the medieval engineers, then there are two enhancements we can make to our passing skills. First, add a quick flick at the end of the passing motion. Instead of lamely releasing the ball, bolt on a last-second flick of the wrist to force the ball out. Your wrists, then, are like the sling at the end of the trebuchet.

Second, move your body like a counterweight to add momentum. This is a bit like radioactivity. When an atom decays radioactively, it sends out, say, an alpha particle. The momentum of all the bits of the atom must be the same before as after. So, if the mass of the nucleus is M and the alpha particle has mass m and is ejected with a speed v, then — ignoring relativity and quantum mechanics — the atom recoils at speed u where $u = mv/M$. This 'radioactive passing' is the key to success. In any game, the backs will swing their bodies one way while passing the ball the other way. Without realizing it, the centre imitates the radioactive nucleus of a uranium atom as it emits an alpha particle. If Castaignede — who was 84 kg in his playing days — could swing back at a tiny speed of 10 cm/s, the 425 gram game ball rockets out at about (84 x 0.1)/(0.425) ~ 20 m/s, which if sent out at the optimal 45 degree angle will travel 40 metres at a speed so great the enemy backs can't intercept it – unless they encroach offside. What's more Thomas, unlike Lamb Chop, is still on his feet.

SCORING IN THE CORNER

Sprinting unstoppably, changing direction randomly, and passing the ball firmly are all wonderful things. The greatest players in the world can do all of these things, and more, without much effort. And yet, even they sometimes have to dive for the line and score in the corner. If you end up short of the line but your momentum carries you over, you've just arrived at Five Points land. If you have to worm your way over, the referee blows up and you get grief from your comrades for having given up a penalty.

So, how far out can you dive and score? Newton, again, comes to our rescue. Suppose you make that last-gasp game-saving attempt by heroically launching yourself at a speed v at an angle of 45 degrees to the ground – the best angle to maximise your dive. You'll hit terra firma again after you've travelled v^2 / g. If your horizontal speed is not affected by making contact, it remains unchanged at $v \cos 45 = v / \sqrt{2}$. Now the ground starts to slow you down. Suppose the *dynamic* (not static) coefficient of friction between you and the ground is f. The deceleration you experience is fg. For this constant acceleration, Newton predicts you'll come to a halt after travelling a distance:

$$s = \frac{v^2}{4fg}$$

If we add this to the distance you'll already covered through the air, you can dive for the line and score from a distance:

$$D = \frac{v^2}{g}\left(1 + \frac{1}{4f}\right)$$

So, if you run at 10 m/s on a completely non-slippery surface, you can score from 12.5 metres out. If you roll along at a mere 5 m/s, though, you'll be able to cross from only 3.125 metres out. The smaller the value of friction, the longer you can slide: diving over on a muddy day works far better than trying to slide over a bone dry surface.

Bear in mind this is sliding friction, used for two dry surfaces. If it gets very muddy, you become a solid body moving over something that is a liquid. A plane or a car, if they travel at too great a speed can hydroplane. Hydroplaning is a complicated subject – NASA has a 74 page report on it. On the bright side, above a certain speed you will no longer grind along in contact with the turf, but glide along on a jet of water and mud combined, allowing you to slide far farther. For a plane hydroplaning on the tarmac, engineers sometimes use $f = 0.05$. If the same were true for humans, you could dive for the line at 10 m/s, you could slide over for the momentum try from more than 60 metres out – wouldn't that be nice! Alas, planes landing on tarmac are far different from humans thumping onto turf.

In one of the games I will always remember, I was drafted in to play inside centre. I'd always played in the front row but, as we'd always comfortably beaten the team we were up against, our coach thought it would be fun to stir the mixture a little bit. Unfortunately, a lifetime spent playing in the front row doesn't really prepare you to do anything mentioned in this chapter: I could sprint a bit, it's true, but I could spin pass only to one side and had no sidestep worthy of mention. So, that afternoon, whenever I got the ball, I ran straight for the line. I couldn't easily pass, nor jink around the pitch; all they had to do was stop me. I went over for four tries before the rain started. Within a few minutes, water covered the surface. I got the ball again from a few yards out and dived for the line. Luckily, the coefficient of friction had been lowered so, even though I landed a couple of metres short, I slid over the line. It was a great feeling. My opposite number was not pleased. The second I stopped, he whacked me on the back of the head, so that my face was covered, and my mouth was full of mud. It was worth it. In spite of scoring five tries, I was never asked to play centre again. I guess my efforts in the front row were worth more than five tries to my high-school team!

Spare a thought, though, for the teams on the losing end. The worst spanking in international rugby (so far) came on October 27, 1994. In case you missed it, Hong Kong took on Singapore in Kuala Lumpur, a city famous for hosting the 'Rumble in the Jungle' between boxers Mohammed Ali and Smokin' Joe Frazier. In 1994, the only rumbling came from Hong Kong, for they won by an astounding 164-13, with 26 tries to their credit. As 17 were converted, I suspect that many of them weren't scored right in the corner. Singapore, 3 years and 364 days later, took on India and beat them 85-0. You win some, you lose some…

CHAPTER TWO - ENDNOTES

1 Hill's equation has $R = kv$. Therefore, $m\frac{dv}{dt} = F_{MAX} - kv$. This gives:

$$\frac{dv}{F_{MAX} - kv} = \frac{dt}{m}.$$

The right-hand side is the derivative of *t/m*. The left-hand side is the derivative of

$$-\frac{1}{k}\ln(F_{MAX} - kv).$$

Combining the two means that:

$$\ln(F_{MAX} - kv) = -\frac{kt}{m} + const.$$

which can be written:

$$F_{MAX} - kv = A\exp(-kt/m),$$

where A is a constant. When t = 0, v =0, so $A = F_{MAX}$. Writing $k = F_{MAX}/u$ allows us to rearrange the mathematical furniture and set $v = u\left[1 - \exp\left(-\frac{F_{MAX}t}{mu}\right)\right]$.

2 For fun, we can provide a plausibility argument using a lovely piece of science known as dimensional analysis. The time is measured in seconds. Whatever the true formula is, we know the right hand side of the equation must also have the units of seconds. The relevant items are that the length of the pendulum, *L*, is measured in metres. The mass of the pendulum is *M*, measured in kilograms, and the only force acting is gravity, so we need to worry about the acceleration due to gravity, *g*, which has units metres/square second. Somehow we need to combine *L*, *g*, and *M* so that we have the units of seconds. Form a quantity $M^aL^bg^c$. The units of this strange beast are: $[kg]^a[m]^b[m/s^2]^c = [kg]^a[m]^{b+c}[s]^{-2c}$. If the units are seconds, then we must have $-2c$=1, so that c = -0.5. Kilograms can't enter into this, so a=0. Neither can length, so $b + c = b - 0.5 = 0$, so that b= ½. Putting it all together, we must have $T \propto \sqrt{L/g}$.

3 For a fair coin, after *N* sidesteps, Phil has gone Northwest *H* times – the number of heads he's tossed. He has gone Northeast exactly *N-H* times. As each sidestep Northwest takes him 1 metre left, Phil is a total distance of $(N - 2H)$ to the right of the centre line. If we toss the coin *N* times, we expect that it'll come down heads

half the time, so that the average number of heads is *N/2*. In that case, Phil's average distance from the centreline [the average of $(N-2H)$] is 0.

His (unaveraged) distance from the centreline is:

$$R = S_1 + S_2 + \ldots + S_{14}$$

If the coin toss on his 6th sidestep is heads, then $S_6 = -1$ and if the coin toss on his 11th sidestep is tails, then $S_{11} = +1$. The square of Phil's distance from the centreline is:

$$R^2 = (S_1^2 + S_2^2 + \ldots + S_{14}^2) + 2(S_1S_2 + S_1S_3 + \ldots + S_1S_{14}) + 2(S_2S_3 + S_2S_4 + \ldots + S_2S_{14}) + \ldots$$

Since the square of -1 and +1 are both 1, the first term in parentheses is 14, the average of which is, of course, 14. Consider all the nasty terms, such as S_2S_7, which are the products of two different S values. If both of the S's are 1, or both are -1, then their product is also 1. If one of them is +1 and the other -1, the product is -1. This exhausts the possibilities and so, on average, the product adds up to 0. So, the mean square distance is:

$$< R^2 >= 14$$

or

$$\sqrt{< R^2 >} = 3.74$$

4 M.P. Silverman, Wayne Strange, Chris Silverman, and T.C. Lipscombe, 'On the run: Unexpected outcomes of random events' *Physics Teacher*, 37 (1999): 218 - 225.

5 See, for example, Don S. Lemons and T.C. Lipscombe, 'The Shape of a Randomly Lying Cord', *American Journal of Physics* 570-574 (2002). The total distance *R* travelled by a player who ricochets off opponents obeys $< R^2 >= 2v^2T^2\left(\frac{t}{T} + e^{-\frac{t}{T}} - 1\right)$. Here *T* is some characteristic time. When $t << T$, the short-time limit is $< R^2 >\approx v^2t^2$, so you run in a straight line. After a long time, when $t >> T$, we have $< R^2 >\approx 2v^2\, t/T$, which has the same form as the usual Brownian-motion equation.

6 He derived an equation to describe the probability *p(x,t)* that a particle initially at a position *x*=0 at time *t*=0 would be at the point *x* at a later time *t*. It can be obtained from our simple random-walk model, but only after some intense mathematical manipulations. The main result was that

$$p(x,t) = \frac{\exp(-x^2 \frac{svt}{3} \sin A)}{\frac{4sv\pi t}{3} \sin A}$$

This describes a gas where a molecule travels, on average, a distance $s \sin A$ before colliding with another, and in which a molecule travels with speed *v*. In our language, it is the probability that Phil is a distance *x* from the centre of the field, given that he runs with speed *v*, has a stride length of *s*, and dodges either to the right or left at an angle *A*.

7 The kinetic energy *K* of a body of mass *m* moving with speed *v* is $K = mv^2/2$. The rate of change of this energy is

$$\frac{dK}{dt} = \frac{d}{dt}(\frac{1}{2}mv^2) = mv\frac{dv}{dt}$$

8 The Hill-Keller model is that the athlete's energy *E(t)* obeys:

$$E(t) = E(0) + St - \frac{1}{2}v^2(t) - \frac{k}{m}\int_0^t v^2(t)dt$$

9 The formula for the range is:

$$R = \frac{3(1+2x)L \cos A \sin 2A}{(1+3x)}$$

This is a constant, which need not worry us, multiplied by the expression $f(A) = \cos A \sin 2A$ which, expanding the double angle, is $f(A) = 2\cos^2 A \sin A$. The derivative is:

$$\frac{df}{dA} = 2\frac{d}{dA}\left(\cos^2 A \sin A\right) = -4\cos A \sin^2 A + 2\cos^3 A$$

This is zero when the angle *A* is:

$$\cos^2 A - 2\sin^2 A = 0$$

The solution is $\tan^2 A = 0.5$, which predicts a best launch angle of *A*=35 degrees.

Chapter Three

Crunch Time: On chasing, tackling and injuries

'Bloody typical, isn't it? The car's a write-off. The tanker's a write-off. But JPR comes out of it all in one piece.'

Gareth Edwards, commenting on JPR's traffic accident.

'You've got to get your first tackle in early, even if it's late.'

Ray Gravell

Before you can tackle Harry Ellis, you have to catch him. That's far easier said than done, but in the world of mathematics – and sometimes in reality – all things are possible. Suppose that the Gloucester prop Christian Califano, the bad boy of French rugby, has to chase Ellis down. How can a slow-moving prop catch a swift back? Ellis can sprint at high speed; Califano cannot. The answer lies in something every rugby coach knows and every back must learn: run straight.

Suppose Ellis and Califano are at the same ruck at the same time. The ball comes out safely in Ellis's hands and he runs downfield. Every step he takes is quicker than the Toulouse player, but Ellis might have to weave his way through a morass of French defenders. Califano must use his brain – not something props are noted for, I grant you – but may see a gap in the French defense, one that he can plug. Ignoring Ellis's current position, the prop heads for the opening. He arrives, just in time to meet and greet the Leicester fly half with the customary French hospitality.

The physics that explains how this is possible is one word: vectors. Vectors have magnitude and direction. Ellis and Califano start off at the same point (the ruck) at the same time. They end up at the same position (the tackle) at the same time. (See Figure 3.1) Ellis has run a far longer path than Califano. Perhaps he moved according to the coin-tossing model or the bouncing-off-of-players model of running from the last chapter. No matter what, he has run a far greater distance. He has run every step of it faster than Califano can. But when you add all the velocity vectors at each moment of Ellis's run for glory, they combine to give a single resultant vector that is exactly the same in direction and length as Califano. While Christian is slower every step of the way, he and Harry have the same average velocity! This provides a clue: as they say on the Tube, 'Mind the Gap' — if you're not that fast, ignore the ball carrier and go stick your fingers in the hole of your defensive dyke by plugging the nearest opening in your defence. The ball carrier might obligingly run into you.

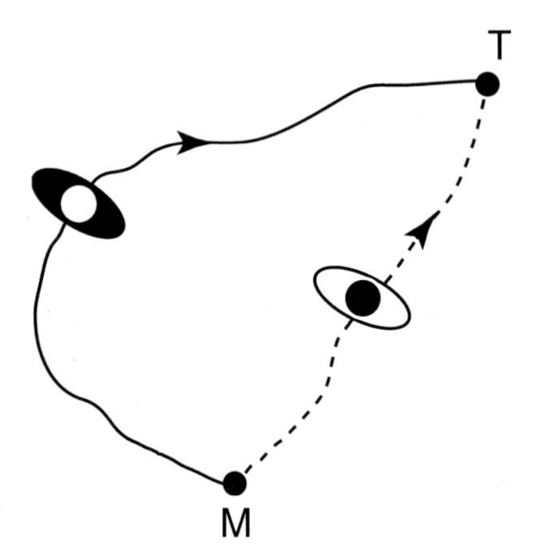

Figure 3.1: Ellis (solid line) runs with the ball from a maul (M) but takes a longer path than the slower Califano (dotted line) but they both meet at the tackle, (T). Their average speeds differ, but their average velocities are equal.

For Califano to catch Ellis, the prop pretty much has to ignore where the scrum half is *now*, guessing where he *will be* at some stage in the future. Backs don't have that luxury. They have to sprint after the person they are marking, or speed off to cover a back if an attacker has broken through. But as with Christian, life is easier if you know where the attacker is headed.

TRIVIAL PURSUITS

Rugby is a thrilling game but sometimes it's obvious what's going to happen next. If the Italian fullback David Bortolussi joins in their line but the

Azzurri somehow cough the ball up, the Irish may well thump the ball down the pitch into a central spot beyond the Italian tryline. The race is now on. Ireland winger Tommy Bowe has *R* metres to go to touch the ball down and score the try between the posts. Bortolussi has *L* metres to get there first and concede a five metre scrum. Who wins? If Bowe sprints at *v* m/s while David manages *kv* m/s, the Italian wins win the foot race if $L/R < k$. Physics and real life coincide once again: the faster you are (the bigger *k* is) and the farther away you can be and still save the game. If a prop lumbers through your line at 5 m/s and is 10 metres from your line, you can stop him if you can run 10 m/s and you are less than 20 metres from him. The lesson is that the faster you are (to increase *k*) and the better you play the position (to decrease *L*) the fewer points your opponents will score. When you have the ball, there's role reversal. If you are 10 metres away from the line with only their prop to stop you, he'll have to be less than 5 metres out or else you'll ground the ball without much effort.

Califano may trundle around the pitch trying to plug a gap into which an opponent may blunder. With a kick ahead, or when a lone breakaway runner heads towards the posts, you know where to go and can move there at flank speed. These situations are easy. For most of the game, though, when the ball comes out on their side of the scrum and your opposite number takes possession, you're never quite sure what to do. (Speaking of opposite numbers, the All Blacks were the first to use numbers, in 1897. Wales used letters as late as 1936; Leicester and Bristol used letters until the rules changed, while Richmond and Bath didn't use the number 13, causing much confusion for their opponents). But whether your enemy uses numbers or letters, the course of action seems straight forward: head towards him now, as quickly as you can, then tackle fiercely.

One of the greatest tacklers in modern rugby is Allan Bateman. He hails from the tough town of Caerau in South Wales and began playing for his local club Maesteg — the Old Parish — before going on to play and star for Wales in rugby union and rugby league. So ferocious and skillful was his tackling that Allan was nicknamed 'The Clamp'. Once Bateman got his pincer-like arms around you, there was no way you could remain upright. He was one of the few players ever to prevent Jonah Lomu, the leading try scorer in RWC history, from scoring a try in an international rugby match.

Suppose Allan is in search-and-destroy mode. Jonah has the ball, but not for long. Bateman starts to run, at constant speed, to where the All Black is right now. One step later, Allan is a tad closer to Lomu, but Jonah has also moved slightly. So, Allan steers a new path, to where Lomu is right now. This game of cat and mouse continues. What happens?

THE WAY TO GO: PURSUIT CURVES

Geometry, the branch of mathematics that describes Bateman's movements, tells us about pursuit curves. The details were worked out by two scientists from a couple of Six Nations countries, Pierre Bouguer of France and George Boole of England, who spent much of his life in Ireland. Boole's wife was the niece of Sir George Everest, after whom the mountain was named.

Pursuit curves can become nasty, mathematically speaking. For a computer, they are relatively simple. Start off by labelling the pitch. The centre of the pitch is *Y=0, X=0.* Life is sweeter, at least in terms of mathematics, if Lomu gets the ball smack in the middle of the pitch, so that the Kiwi starts off at (0,0). He wants to score between the posts – who doesn't – so he runs at full tilt, v m/s, straight down the pitch. His X co-ordinate is always going to be 0. After t seconds, the man in black is at the position $(vt, 0)$.

So much for the New Zealander. When Lomu gets the ball, the Clamp is at some position on the field, say (y_0, x_0). If y_0 and x_0 are positive, then he's in his own half of the field somewhere over on the right-hand side – as Lomu looks at him. Bateman now heads to where Jonah is, and Allan has a position $(y(t), x(t))$.

This, combined with the fact that Bateman's speed is a constant kv, is enough to work out the path along which the Clamp runs[1]. It's complicated, to say the least, but it makes a fascinating prediction. Namely, suppose Allan starts off in his own half running directly towards Lomu (See Figure 3.2).

The Welsh warrior will reach a certain point when he's running straight towards the touchline, before he curves around and starts running back towards his own tryline (see Fig. 3.2). That point is when his distance from the centre of the field is:

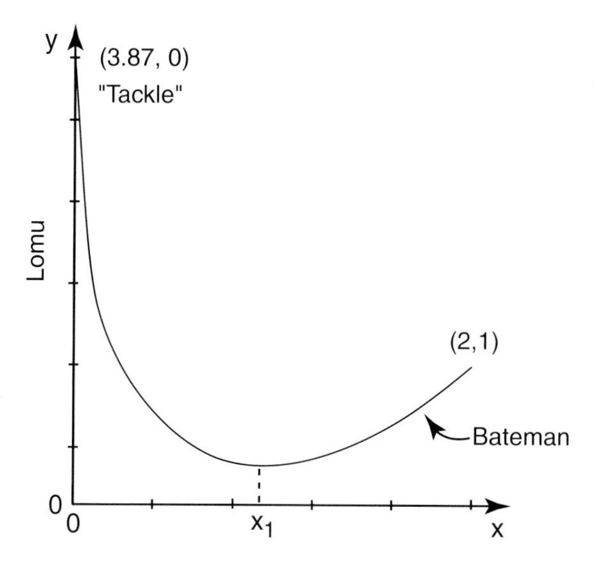

Figure 3.2: Bateman starts at (2,1) and chases Lomu who starts at (0,0). If the relative speed is 1.5, Bateman tackles him at (3.87,0).

$$x_1 = \frac{x_0}{\left(\frac{y_0}{x_0} + \sqrt{1 + \left(\frac{y_0}{x_0}\right)^2}\right)^k}$$

Again, this corresponds to a rugby fact of life. You'll often see a player head downfield towards his opposite number, only to see him have to change direction. This wasn't, after all, misjudgment. Instead, he's doing exactly what he's paid to do – keep track of the danger by sticking close to the man he's marking.

Some complicated sums shows that Jonah tastes turf when the tackle transpires at:

$$y_{tackle} = y_0 - \frac{1}{2x_1}\left[\frac{\left(\frac{x_0}{x_1}\right)^{1+\frac{1}{k}}}{1+\frac{1}{k}} - \frac{\left(\frac{x_0}{x_1}\right)^{1-\frac{1}{k}}}{1-\frac{1}{k}}\right]$$

There is a problem when runner and tackler both have the same speed. In that case, $k = 1$ and so y_{tackle} becomes infinite. Put less mathematically, Bateman can't catch the All Black and the 'head where the player is now' strategy won't work.

If Lomu is obviously headed straight down the pitch towards the goal line, it's clear where he's headed and you can run to where you hope he'll be. But if it's not clear where he's headed and you run towards his current position, you may well never catch up with him. Physics does have a neat solution to the problem, but sad to say it's difficult to put into action.

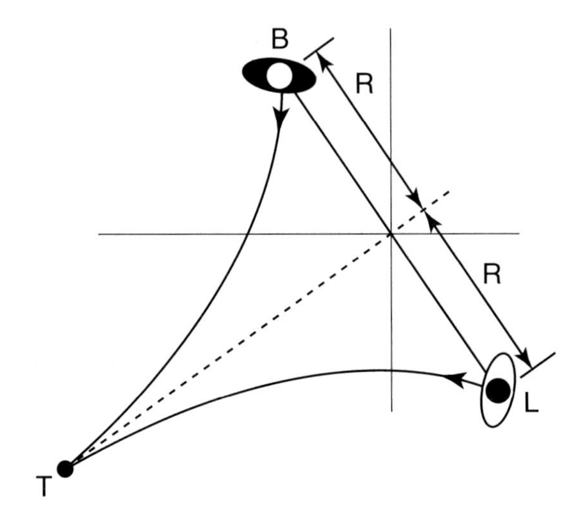

Figure 3.3: Bateman (B) tackles Lomu (L) by mirroring his moves.

The moment Jonah gets the ball in his hands, imagine a line drawn between the All Black and Bateman. Half way along this line, draw another line perpendicular to it. These two lines are going to be the new coordinate system. (See Figure 3.3) All Bateman has to do is mirror the New Zealander. That is to say, if Lomu moves in the direction of the perpendicular at speed v, Bateman does so as well. But if Lomu moves 'across' the perpendicular at speed u, Bateman runs with speed $-u$. If the two players are originally separated by a distance $2R$, they will collide in a tackle at a time $T = R/u$. By then, they'll have moved a distance $d = vT$ down the perpendicular, so that $d = vR/u$. If the line between Lomu and Bateman is at an angle A to the usual axes (down and across the field), then the farthest downfield that Lomu can get before being tackled is $(vR/u)\cos A$. This appeals to common sense, for the farther away the players are to begin with, the further Lomu can run before getting caught. Lomu's best strategy, it turns out, is to work out where the perpendicular will intersect the touchline and head toward this point, for this will maximise how far he'll get downfield before the Clamp is put into action. Unlike the 'head in the direction of your opponent' strategy, this mirroring strategy will always result in a tackle; whether it's too late to save the try is another matter entirely![2]

DOGGED DEFENCE

There's a lovely feeling when you break through an opponent's line. There you are, wind flowing through the hair, the try line in front of you. What could be better? Well, knowing no-one can catch you… It's truly disheartening when you are on defense and you know you can't catch the guy with the ball. To make matters worse, he can slow down, knowing full well that it won't matter. All he needs is a little comfort room, something big enough that you can't dive and trip him up. Then, as he puts the ball down for the score, he's used the least amount of energy possible and gets to see you sprawled on the grass. For him, it won't get much better.

Pursuit curves provide the model once again. As before, you are condemned to sprint as hard as you can, always following the chap with the ball. His response is to slow up, making sure that you are always a certain length L away from him – too large to cover in one desperate dive in the shadow of the posts.

Their winger toys with you. He, like Shane, has unrivalled speed. He starts off, as before, in the middle of the pitch. To make life easier, suppose you start off on the half-way line as well, but a distance L to one side. The distance is only just too far for you to dive and reach him, so you set off in hot pursuit. The Shane Williams wannabe runs down the pitch, again making sure that his X coordinate is forever 0 and keeping you forever the same tantalizingly close distance behind. That's all we need to predict how you'll move. You'll describe the curve[3]:

$$y = L\ln\left(\frac{L+\sqrt{L^2-x^2}}{x}\right)-\sqrt{L^2-x^2}$$

(See Figure 3.4) The embarrassing thing about this curve is its name. Not its usual name – the tractrix – for that comes from the Latin verb trahere, which means to pull. In German, though, it's the Hundkurve – the hound curve. That's because it was originally used to describe what happens when a dog's master walks along the y axis pulling little Fido behind him, making sure that the yappy little poodle doesn't come close enough to bite. In that case, *L* is the length of the dog's lead. Yes, not only does the winger get the points and get to see you dive in the mud behind him, he also gets to relish you following in his footsteps like a naughty poodle. And, to make matters worse, he can run as slowly as he likes, keeping you out of tackle range, so that he is as fresh as a daisy next time he gets the ball – and so can steal past you yet again.

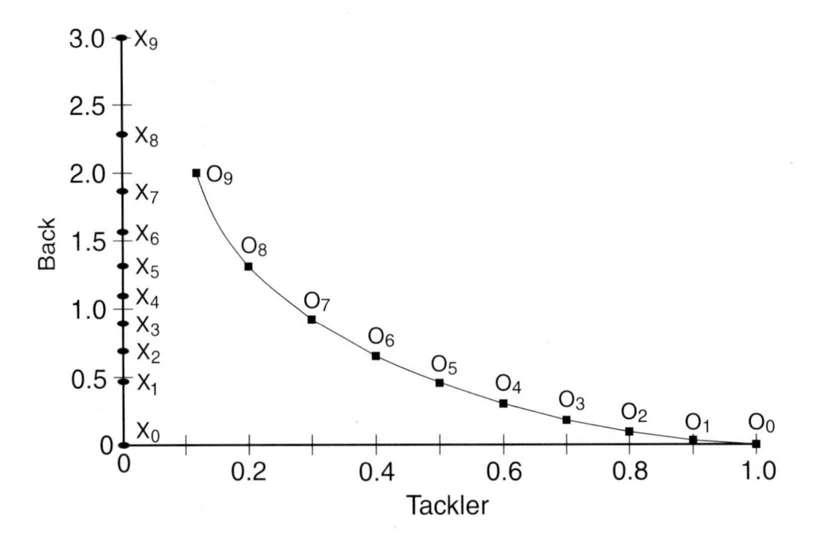

Figure 3.4: The Tractrix: A speedy back (X) remains a constant distance from the tackler (O). The position of each player is shown by the subnumeral at time t_0, t_1, t_2, etc.

GETTING TACKLED

In the great game of rugby, more often than not, you seldom get to run in the open, hair flowing gently in the breeze. Usually, within a few steps of gaining control of the ball, someone is on you, trying to deck you, rip the ball out of your arms, or both. Sidestepping, sprinting as fast as you can, may gain a few more metres but, in a sense, you are almost certainly doomed to failure. If you have possession of the ball long enough, you will nearly always get tackled.

There is some off-the-shelf physics that can explain why it is so easy to be tackled and how hard it is to score. The physics was developed to explain how molecules move inside a gas – the same branch of physics that results in the random-walk model. As you run along, ball safely grasped in both hands, you have a speed v. You also have a certain width, label it w. Suppose every member of the defending team can hurl themselves a distance R and tackle you successfully. This means anyone to your left or right who is within the distance $R + w/2$ can bring you down — or at least slow you down. What can we learn from this?

Every second, you trace out a 'tackle tunnel' whose area is $v(2R + w)$ (See Figure 3.5).

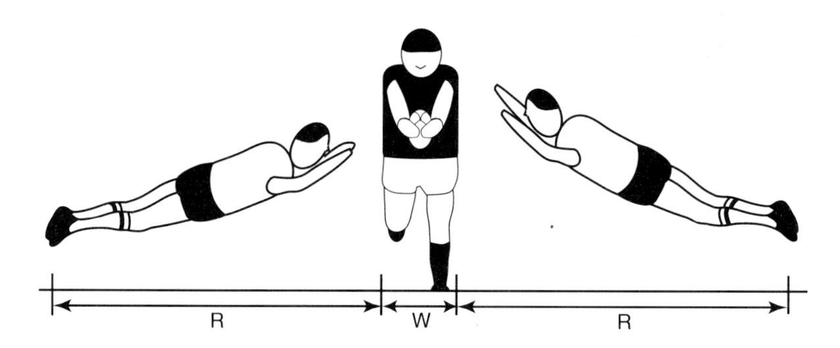

Figure 3.5: The tackle tunnel: Anyone within (2R+W) of you can tackle you.

The first rule here is that the smaller the area, the harder you are to stop. So, running with an arm flailing is a bad idea, for it greatly increases w. The second lesson, when you're doing the tackling, is that the farther you can dive, the more people you can tackle, for your value of R is high. Develop those thigh muscles, quads, and the rest so you can leap and lunge onto your opponent.

The opposition has N players on the pitch, all of whom bay for your blood. This converts into, say, n opponents per unit area of the pitch. In a time t, on average, the number of tackles you'll encounter is the area of the tackle tunnel you have carved out multiplied by the number of opponents per unit area. This is $nvt(2R+w)$. Put another way, on average you have to fend off one tackle every $1/nv(2R+w)$ seconds, and you can expect to run a distance $1/[n(2R+w)]$ before being tackled. This, in the world of molecular dynamics, is the mean free path of a molecule – how far it can go before bumping into another one. Technically speaking, we ask that the tackle tunnels don't overlap. This means that the physics equivalent is a dilute or low-density gas.

Even though Scotland's defense looks static, it can sometimes move about a bit when push comes to shove. Our description, though, has a player running full pelt into a static opposition. A high-falutin' theoretical model accounting for opponents that move adds a factor of $\sqrt{\pi/32}$, about 1/3, meaning you can hope to go the distance:

$$\sqrt{\frac{\pi}{32}}\left(\frac{1}{n(2R+w)}\right)$$

before tasting turf. This highbrow model is also based, like Phil Bennett's running, on a random walk model. In other words, our player's aren't too swift (in terms of brain speed, not running speed) and so don't react to the play that's going on around them. They plod, zombie like, across the pitch making left-right choices at random. Still, the results look good — even though our gas has only 15 molecules, not the millions it ought to have for the theory to hold.

If you want to defend successfully, behave. Players sent to the sin bin decrease n. So, make sure your team is in a perfectly equitable mood before the day's festivities begin. Captains and pack leaders should try to keep everyone reasonably calm during the game, even if the other team is being a tad naughty.

Second, make sure you are fit. If your 15 players are scattered over the pitch, whose area lest we forget is 69 metres by 100 metres, then $n = 15/(69 \times 100) = 0.0022$ players per square metre. If so, when your enemy has the ball in your 22, then n is 0.0022 and only three of your players (15 x 22/100) are onside and ready for action. In contrast, in a fit team all fifteen players are onside and within your 22, eagerly planning to tackle the opposition. If so, you cram all 15 into an area 69 x 22, so that n is a healthy 15/(69 x 22) = 0.01, five times higher than before. It turns out that 69/15 is 4.6 so that, if the 15-man defence is stretched across the field, each defender has a column 4.6 metres wide and 22 long to protect. Your stalwart defenders, all of whom can dive 2 metres and have a width of one metre, therefore block the entire width of the pitch and are guaranteed to tackle the ball carrier at least once before he can get to the line.

The situation where everyone remains onside is similar to the world of thermodynamics, where a piston compresses a gas. Here, the piston is the offside line, usually determined by where the ball happens to be. The gas is the players in the defending team. As the piston moves in – the ball carrier moves forward – the gas is compressed, and its density increases. (See Figure 3.6). The closer you get to the opposing tryline, the more you compress the defence, and foreign jerseys seem to be everywhere. In fact, if you get the ball at your opponent's 22 and all of them are onside, there's a 50% chance that one of them is within 50 square metres of you, which amounts to a 50-50 chance that someone is within a radius of 5 metres waiting to tackle you once you obtain the ball. In real life, the opposition is not scattered randomly within their own 22, so the chances are far higher that one of them is within 5 metres – tackling distance – of you.

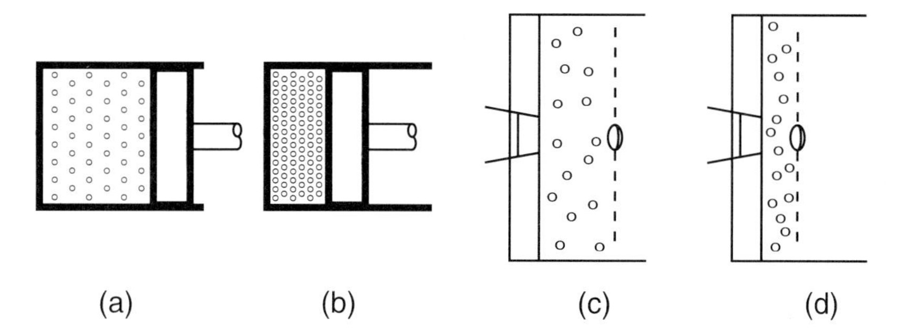

Figure 3.6: A piston compresses a gas [(a) and (b)] like the offside line compresses the defense [(c) and (d)].

Putting in some numbers, if you pick up the ball inside your own half, you might get to run about 11 metres before getting tackled. That assumes that you are about 1 metre wide, that players can dive and tackle a player 3 metres away. If you pick up the ball only ten metres out, expect to go only about 1.5 metres before being tackled. What the equation suggests, and experience confirms, is that it is difficult to score a try against an exceptionally fit team – especially as our equations apply to a randomly moving defence, not one that homes in on whoever the ball carrier may be.

We can do better. Suppose your chance of getting tackled in your next step is some number p. We don't know, yet, what p is. All we ask is that this probability is tiny and doesn't change from step to step. The chance of not being tackled after two steps is the chance of *not* being tackled on your first step $(1-p)$, nor on your second. You emerge unscathed after two steps with a probability that is $(1-p)^2$. In general, you still hold the ball after n steps with a probability $P(n)$, which is $(1-p)^n$. This expression for the probability occurs in many diverse areas of life. For example, if your car depreciates in value by a factor of p each year, then after n years your car is worth only $(1-p)^n$ of what you paid for it. Also, if p is the chance that a back drops the ball or messes up the pass, then $(1-p)^n$ is the probability of stringing together n successful passes. If there's a 50-50 chance that a back will mess up on any given pass, the chance that the ball will go safely from scrum half to fly half, then on to the inside and outside centre and successfully into the hands of the winger is about 3 times for every 100 attempts. No wonder it doesn't happen that often!

Returning to tackling, to model p it's reasonable to suppose that the longer your step – the more ground you cover – the more likely you are to be tackled. The n th step takes you a small distance dx. It changes your probability of being caught by a small amount, dP. Calculus[4] ordains that the chance of running a distance x before being tackled is:

$$P(x) = \frac{1}{a}\exp(-\frac{x}{a})$$

This is called, for obvious reasons, the exponential distribution function. It is a close relative of the Poisson distribution, which is used to describe rarely occurring events including, famously, the number of Prussian cavalry officers kicked to death by their horses (which, for the curious, was about 7 per decade.)

We don't know, at least not yet, what the mysterious a is. Mathematics reveals it to be the average value of x. In other words, on average you can expect to grind out a metres before you get tackled. In the language of molecules, a is the mean free path. The exponential function here kills things off pretty quickly. Your chance to get to a distance na without being brutalised is exp $(-n)$. It means that you have about a 33% chance of making it to the average distance. The chance of going twice the distance is feeble — only 13.5%. For those who dream of the long sprint, untouched to the tryline, dream on. A run 5 times the average tackling distance crops up once in a blue moon, about 0.6% of the time. If a is a luxurious 5 metres, then only 6 times out of every thousand attempts can you cross over the try line having picked up the ball at your opponent's twenty two. Physics tells you what you already know: rugby has to be a team sport, where players learn to pass and support each other. The isolated run for glory is just that: isolated.

TACKLING

The Welsh national team is funded by Brains, a Welsh brewery. In France, the Pays-de-Galles boys in red could not wear a jersey that advertised beer, and Welsh fans were eager to see what their heroes would do. That day in the Parc des Princes, the Welsh took the field with the word 'Brawn' on their jerseys, and we all got the message. It's a word that describes tackling - brute strength versus plain brawn, irresistible force against immoveable object. If a large player, mass M and speed V, hurtles towards you — a player of mass m and speed v — the challenge is to hold on for dear life. As long as you put the clamp on, really pincer the player between your two arms, you will have done a good job. Before the tackle, the total momentum of both players was $MV + mv$. Afterwards, as long as you hold on, the momentum is $(M+m)U$, where U is the new speed. From Newton's second (and third) law, we know that momentum is conserved, so that:

$$MV + mv = (M + m)U$$

or

$$U = \frac{MV + mv}{M + m}$$

That's how a tiny player such as Italian winger Pablo Canavosio can help his team if Scotland's monster prop Craig Smith has the ball. Craig tips the scale at 118 kg, whereas Pablo is about 85 kilos dripping wet (which if they're playing at Murrayfield, he may be). If Craig runs at 7.5 m/s and the Azzurro crashes into him at 10 m/s, straight on, then afterwards Craig will still move forwards, but at a tiny 17 centimetres every second. The little man has brought the big one to a screeching halt and help – the Italians hope – will not be far behind. David, physics says, can beat Goliath; the charge of the Heavy Brigade can be halted by that of the Light Brigade. All you need to do is play rugby hard from the first whistle to the last, blasting in to every tackle at top speed.

In real-life match situations, it's not that common to have a head-on tackle. It can happen sometimes in a penalty situation near a tryline. I was once playing a game that featured a giant on the opposition. We were losing by a hatful, and he, their pack leader, kept shouting 'tall and proud' at the lineouts. I exchanged a few remarks with him and suggested that he might care to cease such activities. Soon there was a penalty close to our line. Their plan, which was obvious, was to give him the ball at full steam. As soon as their scrum half tapped the ball, I sprinted straight for the second row and clamped my arms around his legs. My mass was far less than his but I could sprint quickly. My speed was far higher than his, the laws of physics were with me, and I prevented the try – the concussion was worth it!

The moral is that it's hard to stop players who have a lot of momentum. Momentum is the product of mass and velocity and velocity, so we've seen, depends on the square root of a player's height. One measure of fitness is the body mass index, or BMI. This is your weight divided by the square of your height. Suppose all superfit players have the same low BMI. Then $BMI = W / H^2$ is a constant. Put differently, $H = \sqrt{W / BMI} \propto \sqrt{W}$. The momentum $p = mv \propto (W \times \sqrt{H})$, so that $p \propto W \times W^{\frac{1}{4}} \propto W^{\frac{5}{4}}$. Equivalently, $p \propto H^{\frac{5}{2}}$. This suggests that the tall backs, the ones who are the most massive, are the hardest ones to stop. Again, it's fairly believable, for it accounts for the success of Jonah Lomu (6' 5'), Doug Howlett (6' 1'), and Joost van der Westhuizen (also 6'1'). To explain how come Daisuke Ohata (5' 9') and Shane Williams (5' 7') have compiled so many tries if their momentum is so low, remember that they can accelerate far more rapidly: they're easier to tackle, but far harder to catch!

When you crash into an opponent dead on, with equal and opposite momentum, you end up stationary afterwards. That's the effect of conservation of momentum. What, though, of the kinetic energy? Beforehand, two behemoths have a sizable kinetic energy and afterwards they have none. Also, if they end up on the grass, the two gargantuas have lost potential energy as well.

As energy can neither be created nor destroyed, it can't just evaporate. Some of it goes into sound energy, the 'smack' of the impact that gets the crowd roaring. Part of it goes into compressing the soil on which you land, and another part might be transferred into the kinetic energy of your brain, which makes it collide with the inside of your skull, the event that defines concussion.

If you're in the three-quarters sector of the rugby industry, or if you don't enjoy concussions, you don't typically want to tackle head on. Your plan when defending is to push the opponents out wide, towards the touchline. The enemy has been coached to run straight. When the two of you meet up, momentum is still conserved, but now we need to think about its vector qualities. Their player has a mass M and speed $\vec{V}$, while you have mass m and speed $\vec{v}$. Just as before,

$$\vec{U} = \frac{M\vec{V} + m\vec{v}}{M + m}$$

If he runs down the pitch, $\vec{V} = (V,0)$. I.e., he has no component of his velocity headed across the pitch whatsoever. If you are trying to force him out wide, you might be running at an angle A towards him. 'Towards' signals the presence of a minus sign. So your velocity is going to be $\vec{v} = (-v\cos A, v\sin A)$. Your body, combined with the victim of your tackle, now has the velocity:

$$\vec{U} = \frac{(MV - mv\cos A, mv\sin A)}{M + m}$$

When *A* is 0 degrees, you hit the player straight on, which gives the same result as before. At an angle of *A,* though, there ends up being a velocity component directed towards the touchline. This is useful, for its magnitude is $mv \cos A/(M+m)$ which has nothing to do with how fast the runner is headed downfield when the tackle takes place. If you are sprinting towards him at 10 m/s at an angle of 45 degrees, and you weigh the same as the runner, the sideways velocity will be 3.5 m/s. As long as you are within hailing distance of the touchline, this should be sufficient to earn your side the throw in at the lineout. Again, the physics confirms what experience suggests: hit them sideways on when you are close to the touch line. In a desperation tackle, where the try line is only a few metres behind you, hit them square on with as much speed as you can muster. Ideally, hit them above their centre of mass at a slightly upward angle. That will exert a torque, which will rotate them back towards their own line. It also takes the victim off his feet, so that he can't drive forwards. A classic example was silver-shoed Gavin Henson's unceremonious upending of England's Matthew Tait in the Six Nations game in 2005. Popular on web sites, the famous photo of the Welshman dumping the Englishman a few steps backwards bore the caption 'coming for to carry me home.'

There is a response. As you pound down the five metre line with a score in your sights, don't be put off by the herd and hoard of opponents. Put your arm out for that brush off. If they dive under your arm, they'll be converting some of their sideways momentum into downwards momentum, which makes them crunch into the ground harder and knock you sideways more feebly. You can even step into the tackle, and with a nice sideways swing of the hips you can dispense some sideways momentum in their general direction. And remember — it's always a nice gesture to help them up off the ground once you've scored.

The bigger they are, the harder they fall

If you ever get to carry the ball in a rugby game, the odds that you will be tackled are overwhelming. After all, there are fifteen people trying to catch up with you. Likewise, if the opposition has the ball your job, no matter what position you play, is to stop them from getting any closer to your goal line. The secret, as coaches know, is to wrap up the runner's legs. Rather than half-heartedly slapping at the runner's legs, you really need to use your arms like pincers. With the legs securely locked up, the only place for the runner to go is face down into the mud. If you're being tackled, though, you might want to try to get rid of the ball to a handily placed team mate. The laws of motion give a rough guide to how quickly you'll have to pass while falling.

There are many sturdy individuals who play the game of rugby. Describing them as a solid cylindrical mass of flesh, bone, blood, and muscle might not seem to be praise, but in many ways it is. If our idealised, perfect-world player runs along at speed v, what can happen? The simplest case is a tap tackle. Almost out of reach, the tackler lunges forward and, with a single swipe at boot level, knocks the feet from under the runner. Seen from a physics-eye view, the tap tackle changes the runner's linear momentum into angular momentum. Instead of running down the line, he now rotates about his boots.

We can try a couple of models here. The first is that Jonny collapses like a bag of King Edwards, crumpling on the spot. If so, we claim that his behavior is the same as a particle of mass m that falls from a height $H/2$. That predicts a time to fall of $\sqrt{H/g}$. If Wilkinson is about 2m tall, then the height of his centre of mass above the ground is 1m. On this model, it'll take about 0.45 seconds for him to reach the ground, but this is probably an overestimate.

A competing model is to say that the tackle converts all of his initial horizontal speed into a downward speed. This should provide an underestimate. The time to fall would then be the amount of time it takes his centre of mass to travel the distance $H/2$ at speed v. This is $H/2v$. For a 2 metre-tall person running at 10 m/s, this is about 0.1 seconds.

We expect that the time for him to hit the ground is a few tenths of a second, somewhere between 0.1 and 0.5. That's how long he has to pass the ball or to roll over it and protect it in the ruck. It would be nice to combine both models. One highly suspicious, yet experience-driven suggestion might be as follows. Re-write the H/g term as $H/g = 2H^2/2gH$. Then, to be completely heuristic, note that in gravity problems, it always seems as though motion provides an extra term, so that $2gH \to v^2 + 2gH$. If so,

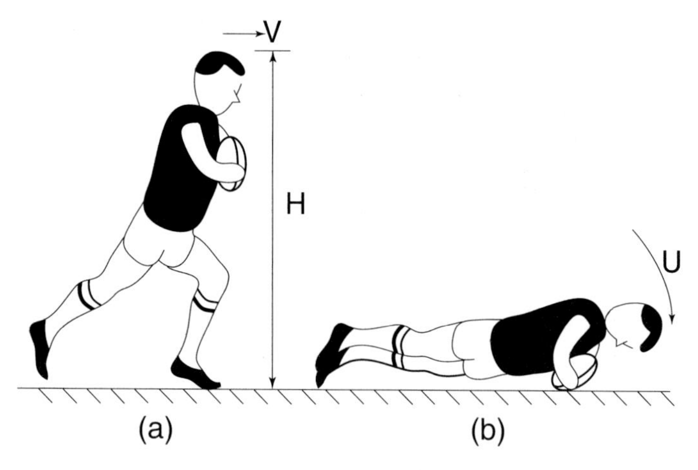

Figure 3.7: Tackling: In (a), he runs with the ball with speed V and height H, then (b) slams into the turf with speed U.

then the time it takes Wilkinson to fall is $T=\sqrt{\dfrac{2H^2}{v^2+2gH}}$. Again, if he runs at 10 m/s and is 2 metres tall, he should fall in a time 0.24 seconds.

Those who frown upon such rough estimates may go into far more detail. If a player behaves rather like a cylinder of flesh, then our tap-tackle model is the same as the rotation of a rigid rod about a point. The exact solution relies on a beautiful, though sadly almost forgotten, set of mathematical functions — elliptic integrals of the first kind. There is a cautionary tale, though. Our model of Jonny Wilkinson as a cylindrical stick rotating about his boots is a reasonable one. It stretches belief, though, to claim that it is an exact one. The high-precision elliptic integrals are probably not warranted; our rough answer is 'good enough.' For example, for our 2 metre, 10m/s runner, the theory predicts a time to fall of 0.35 seconds[5]. You can, though, refine the rotating-rod model. In some ways, the crumpling of a tackled player resembles the collapse of a chimney – a stack of bricks – rather than a rotation. These, too, require a high level of mathematical sophistication, which don't get us much farther in terms of physics.

One thing remains. The time it takes to fall depends on *H*, which we've called the height of the player. In actuality, *H* models the distance of the centre of mass from the bootlaces of the tackled player. If, as you fall, you extend your arms outwards, away from your body, then the effective value of H increases and you will take longer to fall – giving you a split second longer to get the pass off.

There is another way to stave off a tackle. Much of what the Clamp, or any other expert tackler does, depends on friction. Somehow, you have to apply a good firm grip on your opponent's body. Friction, though, has an important role to play. If you can reduce the coefficient of friction of your skin, then a tackler's hand might slide right off you. One way to cut the coefficient of friction is to remove all the surface hair that you have. That's right – shave your legs. Another way is to make sure that your skin is smooth and moist. You can do that by applying moisturizing cream, suntan lotion, or a 'tan in a bottle' to your legs. It sounds silly and not quite the thing a self-respecting rugby player ought to do. But does it work? Certainly. Why else would Gavin Henson do it?

Bone-crunching tackles

At Oxford University, most colleges play rugby games on Wednesday afternoons. In the late afternoon, there is a long queue of people at the John Radcliffe Hospital (the JR2) seeking aid for broken limbs. Tackles, for those who write or speak in clichés, are described as bone crunching and sometimes they truly are. But what, exactly, causes bones to break? The question is important and a simple answer can be found by looking at one of the most humble piece of physics equipment, the spring. If a spring has a spring constant k and you apply a force F to it, the spring extends by a small distance x such that $F = kx$. The energy E stored in this stretched spring is $E = \frac{1}{2}kx^2$. Bones, of course, are not springs. Instead, model them as a column of length L and of some constant (though possibly weirdly shaped) cross sectional area A. The force you need to compress this column by a distance x is, like the spring, $F = (YA/L)x$. Here Y is the so-called Young's modulus, which is a measure of the elasticity of the material in question, in this case, bone. We can use this to express the extension x in terms of the applied force. Thinking in terms of a spring, the spring constant $k = YA/L$.

By analogy, the energy stored in this compressed bone is:

$$E = \frac{1}{2}kx^2 = \frac{1}{2}\left(\frac{YA}{L}\right)\left(\frac{FL}{AY}\right)^2 = \frac{F^2L}{2AY}$$

The last piece of physics to enter the picture is the maximum stress S that a column can take before breaking. This maximum stress, like the Young's modulus, depends on the material you're looking at. This stress, though, is created by the force F. In other words, the greatest force you can apply to the column without it breaking is $F = SA$. Now our work is done. By modelling a bone as a simple column, we predict the energy that is needed to break it is:

$$E = \frac{ALS^2}{2Y}$$

Immediately, we can see this is reasonable. The product AL is the volume of the bone, so that it takes more energy to break a large bone than a small one. The thigh bone is safer than the tibia, for example. The bones of your neck, the cervical vertebrae, are thin and of small area, so these will be prone to snapping should a scrum collapse, as I know from personal experience.

There's nothing much you can do about this situation, in terms of our equation. (Except, perhaps, schedule many fixtures against medical schools. Guy's Hospital claims to have the oldest rugby club in the world). The only medical advice is to prescribe plenty of vitamin D. Milk and sunshine will make sure your bones are thick (enhancing L) and strong (increasing S). Exercises such as yoga, which help increase muscle strength and flexibility may also help, since they reduce the amount of energy being absorbed by your bones. Try to persuade your team captain that a tour of the Bahamas for a Sevens tournament is a good idea – all that sun and relaxation will help you avoid injuries!

People in biomechanics know a lot about the breaking of bones. Car safety depends on knowing things like Y and S for human bones. Odious though it may sound, researchers sometimes use human cadavers in crash tests to see what breaks under what force. They know that Young's modulus for bone is about 1.4×10^{12} Pascals, while the breaking stress is about 10^{10} Pascals. A leg bone might be about 1 metre in length with a radius of about 1.5 cm. In that case, the energy needed to break a leg is about 200 Joules. This is not, in the grand scheme of things, very much. Suppose a large gentleman, of mass 100kg, runs towards you at speed *v.* He tackles but, for some reason, he grabs only one of your legs, which then absorbs all of his kinetic energy, which must be at least 200J. This happens if he comes in at a mere 2 m/s. Fortunately, tacklers normally grab hold of the thigh rather than the lower part of the leg; they grab two legs rather than one; and not all of their kinetic energy is going to be absorbed by the bone itself.

And there's more than one way to break a bone. I was once playing in a game where I was 'slow to get up from a ruck' (a consequence of someone using fisticuffs) on our own 22 and saw one of our team sprinting with the ball on our opponents 22 metre line. He was tackled, and I heard the crack from where I was. He was in the process of side stepping when tackled, and so his feet were held pointing in one direction while he turned his body in the other direction. The tackle clamped his feet, which allowed his torso to exert a torque, and this caused a career-ending spiral fracture of his leg. This gruesome type of injury can also be modelled using elasticity (specifically, the bulk sheer modulus of bone) and the results are the same as for a simple fracture: physics confirms that rugby is a game where bones can, and will, be broken.

NIL DESPERANDUM

Sometimes, in spite of all the training, all the rugby skills, your opponents cross your try line. Never, ever, give up. The rules are quite firm on this – you have to put the ball down on the ground. Crossing over the tryline earns nothing. In a Five Nations game, England versus France, a French winger had crossed over the white paint. His opposite number was Rory Underwood, one of England's most capped players and one of their most prolific try scorers (as someone once wrote, 'imagine how good he'd have been if he'd played for a decent team'). Underwood knew that tackling would serve no purpose so instead he went for the ball. With one mighty swipe, he thumped the ball out of the hands of the ball carrier, resulting in a drop 22 for the Lilywhites. A case of n'allez pas La France. The opposition can help you as well. In a Guinness Premiership game, London Irish faced the Wasps. The ball came through to the Exiles's Juan

Miguel Leguizamon who dived spectacularly over the line, taking one hand off the ball to start celebrating, only to drop the oval object. The Irish ran out winners that day, but how embarrassing. So embarrassing, possibly, that Leguizamon has fled the country, now playing in Paris for Stade Français!

Sevens fans will remember a combination of Rory's ne'er say die and Juan Miguel's brain spasm. In the Twickenham Sevens, when England was up against Wales, Isoa Damudamu crossed the line for the host nation. He eased off the peddle, kissed the ball, slowly lowering it to the ground and savouring the moment, when in came a Welsh defender who booted the ball clear. England fans bay in outrage – are you really allowed to kick the ball from someone's hands, shouldn't it have been a penalty try? – but come on…

THE RUGBY OF PHYSICS

So far, we've looked at the world of rugby and come up with explanations of what goes on, and why, by using the simpler side of science. We can turn the tables, though, and instead of using physics to explain rugby, we can use rugby to explain physics. Rugby, in other words, can provide some lovely similes for what goes on the world of physics.[6]

Take, for example, the scrum. Every player, pretend they are atoms, are packed together in a neat, orderly array. Each one has his predetermined position, and that's where he stays. Any movement is because the scrum, as a whole, is moving. This is rather like the atoms in a crystalline solid. They have a certain pattern in which they are laid out — the face-centred cubic crystal has atoms lined up like oranges in a greengrocers — and when they move, they all move in unison. In a lineout (at least to begin with) and in the line of backs as they wait for the ball to emerge from a scrum, you have a nice orderly string of players, each of whom is about the same distance apart. This resembles the arrangement of molecules in a polymer, the type of chemical from which plastics are made.

The analogy of players to molecules or atoms works pretty well. At a ruck or maul, the players are packed together pretty closely, rather like a scrum. On the other hand, it's a bit chaotic. It's not a regular arrangement; in some parts the players are squeezed together but in other areas there might be a bit of daylight. That's rather like the way atoms are packed in a liquid. The space between atoms in a liquid is roughly the same as that in solids, but the regular pattern is lost. There is, to use a physics phrase, short-range order but no long-range order.

Close to the try line, a maul is a fascinating thing to watch (not if you're playing, for you ought to get stuck in, not stand off gazing). Players join in or break off depending on what the opponents are doing and what side the ball is going to come out on. In other words, players can break off and join the backs. This is similar to evaporation, where the molecules in the surface layers of a liquid (a droplet of water in the bathroom, say) get enough energy to detach from the surface and join the molecules in the gas that surrounds it (the air in the bathroom).

The defence runs around wildly colliding desperately into any opponent who carries the ball. They are like the atoms in a gas, which are high speed and undergo many collisions. Also, the defence is going to be spread out across the pitch, so that the distance between, say, the inside and outside centre is going to be far larger than the distance between the two locks in the scrum. That, too, models a gas, one of whose features is that the atoms are, on average, far farther apart than in a solid.

Last, at least in terms of the properties of materials, we can think about marking. In the wonderful game of winter, you can keep a close eye on your opposite number, but in the end you need to plug a certain gap in the defence. The late great Ray Gravell would not have refused to tackle Jean-Pierre Rives simply because the Frenchman was a flanker. Grav would have dumped Monsieur Rives on his derriere because the Golden Helmet was too near the Welsh line. The need, though, to fill the gaps, to plug the line, to get the person with the ball, is why the model of 'backs as a gas' works well.

In the superb summer game of sevens, something else happens. Sevens is much more man-to-man. If you lose your player, he'll score. In physics, this is rather like an electric dipole. This occurs when a positive charge and a negative charge are placed a certain distance apart and not allowed to move towards or away from each other. Far enough away, the dipole is electrically neutral so none of the other charged particles (the members of both teams) will be attracted to it. As your opposite number goes around the pitch, you move with him, keeping far enough away to keep a good eye on the game, but close enough that you can step up and tackle if he tries any funny stuff. Solids made up of atoms that are bound by electrical forces are enormously difficult to melt. Common or garden salt is a case in point. The sodium atoms desperately need an extra electron and the chlorine atoms have one to spare. You have to blast sodium chloride to over 800 Centigrade or 1,000 Kelvin to get it to melt. It takes that much thermal energy to break the bond. The analogy holds for sevens. If you stick like glue – or like a dipole – to your opposite number, it'll take a huge amount of effort to throw you off. One of my favourite young players, the former Ospreys fly half, once-capped Matthew Jones, found out the hard way. While playing in the Cwmtawe Sevens, the burst of energy he needed to leave his marker behind was sufficient to blow out one of his cruciate ligaments. Matthew needed surgery and the Ospreys, in need of a fly half, signed up James Hook. Mr. Hook now wears Osprey black and Wales scarlet, while Matthew has drifted from the Neath-Swansea club to Moseley, Bridgend, and most recently to the Guinness Premiership team Worcester Warriors. Here, though, Matthew has teamed up with former Wales coach Mike Ruddock, now director of rugby for the English club, and the young fly half has made a promising new start.

ATOM-SMASHING RUGBY

The world of particle physics is full of wonderful, though hideously expensive, toys. Beneath the France-Switzerland border lies the European Centre for Nuclear Research, usually known as CERN. The team at CERN discovered some of the tiny elementary particles that bind together the protons and neutrons in an atom. They found the W and Z bosons and now, in the large hadron collider, hope to find the Higgs boson. Discovering these particles will help verify the picture that theoreticians have painted over the past thirty years, the so-called standard model of particle physics. One of the key tricks in elementary particle physics is to draw simple sketches of how particles interact. Here's where rugby can help.

Represent a player by a dark line and the ball by a dotted one. (See Figure 3.8). The ultimate goal of a back is to catch the ball, run with it, then pass it on. In our diagram the ball is coming towards our player who catches it. He has a slight momentum shift as he 'absorbs' the momentum of the ball. He and the ball run together as one and then, some time later, he passes the ball and congratulates himself on a job well done. To the trained rugby enthusiast, this seems simple enough. To the physics aficionado, it is something beautiful to behold. From a physics perspective, time increases from the bottom to the top of the picture, while distance increases towards the right. The physicist views the illustration as the Feynman diagram of so-called Compton scattering. A physicist can convert those continuous and dotted lines into an equation to be solved that explains the Compton process – how an electron can absorb a photon (a particle of light) and then re-emits it (passes!).

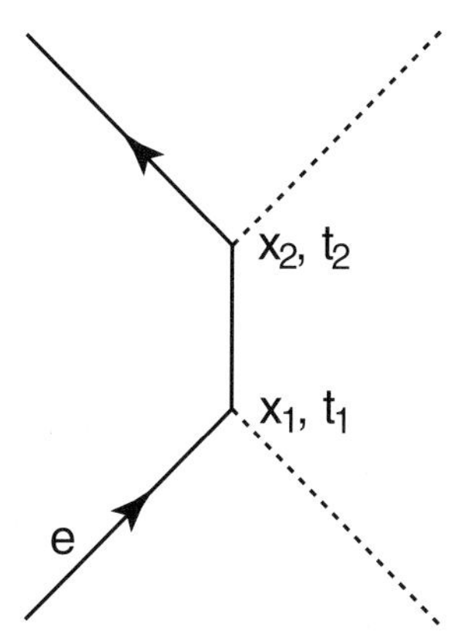

Figure 3.8: Compton scattering: An electron e (dark line) absorbs a photon (broken line) at a point x_1 and time t_1. It emits the photon at a point x_2 and at time t_2. Or; player e catches the ball at x_1, t_1 and and passes it at x_2, t_2.

There's more. In Figure 3.9a, our player P1 has the ball and is headed towards one of his team mates, P2. He passes (emits) the ball at a certain time and soon thereafter our other particle/player catches (absorbs) it. So what? After all, backs do this every day. Well, if we ignore rugby, this is the Feyman diagram explaining how two electrons interact by exchanging a photon, otherwise known as Møller scattering. Here the electrons are our players and the photon – the particle of light – is the ball.

In high-school physics, two electrons repel each other because both possess a negative charge and like charges repel. That's true, but here the repulsion is brought about because of the exchange of a particle, the photon. This is how physicists now look at force: a force is created by the exchange of particles. Who would have thought that a simple pass in rugby could explain the joys of elementary particle physics? You can go even farther. If the receiver of the ball passes it back to the original ball carrier, we have a 'fourth-order Feynman diagram' for electron-electron interactions. (Figure 3.9b)

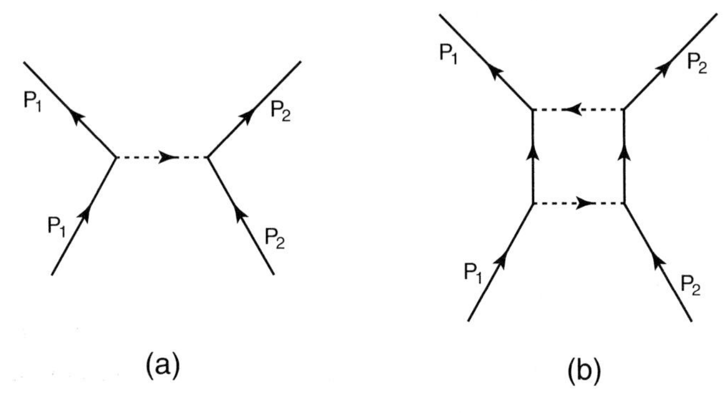

Figure 3.9: (a) Electron-electron (Moller) scattering: An electron P_1 emits a photon that is absorbed by an electron P_2. P_1 and P_2's paths diverge after photon is exchanged. Photon exchange causes same-charge particles to repel each other. Or; P_1 passes the ball to his team mate. (b) A fourth-order Feynman diagram for electron-electron scattering, or the one-two!

It doesn't stop with passing. Take one of the fun moments in playing, when you tackle your opponent and he coughs up the ball. Label your opponent p and yourself $\overline{p}$. You come in, crunch your opposite number, and the ball goes off. The outlook is shown in Figure 3.10, the dotted line going slightly uphill shows that as he was going down, their man made a forward pass – so it's our put in at the scrum. To a physicist, though, it's a Feynman diagram showing a particle colliding with its antiparticle and disappearing into pure energy – a photon. Curiously, to show an antiparticle as a true particle in a diagram, you make it go backwards in time. It might be hard to think of your opposite number moving backwards in time, but think for a moment:

suppose he has the ball and is running for your line. On the video of the game, press rewind. He's now running away from your line, towards his own, which is exactly the way *you'd* run with the ball. So an antiparticle is, in some respects, a regular particle moving backwards in time.

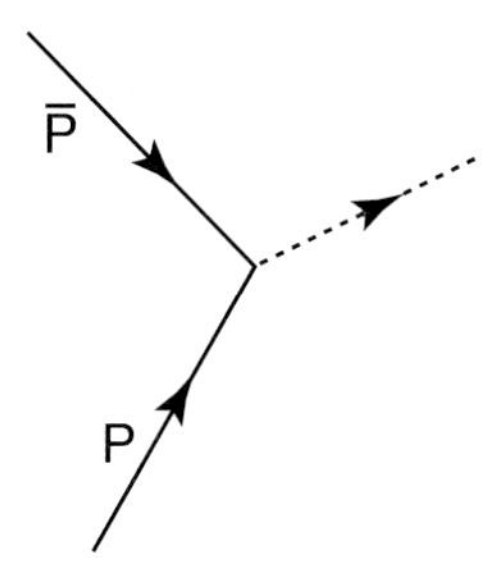

Figure 3.10: Particle-antiparticle annihilation to create a photon. Or; a player collides in a tackle with his opposite number and the ball is knocked on.

Armed with antiparticles, we can re-do our second diagram to show something seriously unpleasant. (See Figure 3.11) To those who study the subatomic world it's an electron-positron interaction, one of the diagrams for so-called Bhabha scattering. For us in the rugby world — it's an interception.

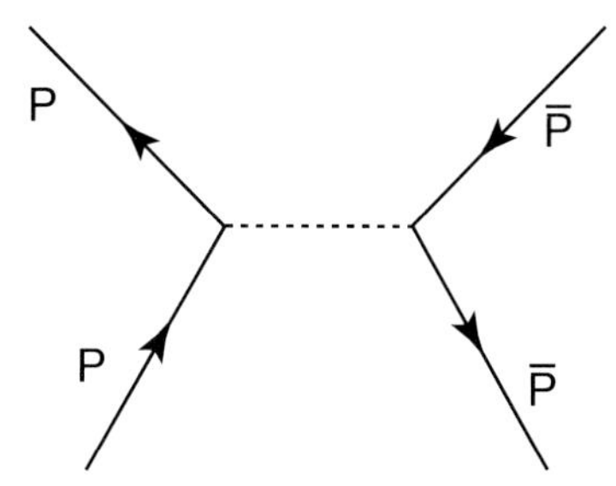

Figure 3.11: Particle-antiparticle interactions. A particle P emits a photon that is absorbed by the antiparticle. Or; an interception.

And last, the universe is full of surprises. A Dutch physicist by the name of Hendrik Casimir showed that a vacuum is not empty; in fact, we get plenty of nothing. Out of what appears to be a vacuum, particles and antiparticles are born and then, as you'd anticipate, collide and die. Place two uncharged metal plates close together in a vacuum and you'd expect nothing to happen, apart from the mild effect of gravity. Casimir tells us that the appearance and disappearance of the virtual particles in the vacuum are messed up by the presence of the plates. The net result is that there's a force, which you can measure. For spectators, rucks and mauls sometimes seem static. Nothing much is going on. Instead, just like the vacuum, it's a seething cauldron where particles, our team, collide with antiparticles, their team, and we almost annihilate each other in a burst of pure energy. OK, it's an analogy. But the next time someone says rugby is not an intelligent sport, tell her it's as easy to understand as quantum electrodynamics!

CHAPTER THREE - ENDNOTES

1 The mathematics here is not for the faint of heart. The angle *A* at which the Welshman steers has a tangent that is the ratio of the difference in their positions (*vt-y*) down the pitch to the difference in their positions across the pitch, *x*. That's to say:

$$\tan A = \frac{vt - y(t)}{x(t)}$$

From calculus, $\tan A = dy/dx$. Therefore $dy/dx = (vt - y)/x$ or equivalently $xdy/dx = y - vt$. The trick now is to differentiate with respect to *x* once again. This gives:

$$\frac{d}{dx}\left(x\frac{dy}{dx}\right) = \frac{dy}{dx} + v\frac{dt}{dx}$$

Use the chain rule to get:

$$\frac{d}{dx}\left(x\frac{dy}{dx}\right) = x\frac{d^2y}{dx^2} + \frac{dy}{dx}$$

and substitute this into the left hand side and cancel to obtain

$$x\frac{d^2y}{dx^2} = v\frac{dt}{dx}$$

Bateman runs at speed *kv*. He travels a distance *ds* in a time *dt*, where *ds=kvdt*. In other words, $dt/dx = (1/kv)(ds/dx)$. The quantity *ds*, though, is the arc length, which is $ds^2 = dx^2 + dy^2 = [1 + (dy/dx)^2]dx^2$. Combining all the expressions together leads to the differential equation:

$$x(d^2y/dx^2) - \frac{1}{k}\left[1 + \left(\frac{dy}{dx}\right)^2\right]^{\frac{1}{2}} = 0$$

To solve this, the substitution *p=dy/dx* helps. We use it to convert the previous equation into:

$$x(dp/dx) = \frac{1}{k}\left[1 + p^2\right]^{\frac{1}{2}}$$

Mathematicians love to solve equations like this, so we can look up the answer they provide, which is:

$$2p = \left(\frac{x}{x_1}\right)^{\frac{1}{k}} - \left(\frac{x}{x_1}\right)^{-\frac{1}{k}}$$

where x_1 is a constant that is determined by the initial position of Bateman.

The next item on the agenda is to integrate the equation once more (as *p=dy/dx*). Again we can look up the solution. If Lomu starts at (0,0) and runs straight down the field while Bateman begins at (y_0, x_0), then Allan's path is given by:

$$y = y_0 + \frac{1}{2x_1}\left[\frac{\left(\frac{x}{x_1}\right)^{1+\frac{1}{k}}}{1+\frac{1}{k}} - \frac{\left(\frac{x}{x_1}\right)^{1-\frac{1}{k}}}{1-\frac{1}{k}}\right] - \frac{1}{2x_1}\left[\frac{\left(\frac{x_0}{x_1}\right)^{1+\frac{1}{k}}}{1+\frac{1}{k}} - \frac{\left(\frac{x_0}{x_1}\right)^{1-\frac{1}{k}}}{1-\frac{1}{k}}\right]$$

where:

$$x_1 = \frac{x_0}{\left[(y_0/x_0) + \sqrt{1+(y_0/x_0)^2}\right]^k}$$

This is the solution to the general curve of pursuit, a result not easily findable in the physics literature. The curve *y(x)* has a minimum at the point $x = x_1$.

As Lomu runs along the *x=0* line, the tackle occurs at a point:

$$y_{tackle} = y_0 - \frac{1}{2x_1}\left[\frac{\left(\frac{x_0}{x_1}\right)^{1+\frac{1}{k}}}{1+\frac{1}{k}} - \frac{\left(\frac{x_0}{x_1}\right)^{1-\frac{1}{k}}}{1-\frac{1}{k}}\right]$$

If Bateman is far quicker than Lomu, the tackle takes place around the point y_0. If Bateman has the same speed as Lomu, no tackle will occur – at least not until well after Lomu has scored a try.

2 See James O'Donnell 'Pursuit and Evasion Strategies in Football', *The Physics Teacher* 33(8), pp516-18 (1995).

3 You are at *(y,x)* while Shane is at *(Y,0)*, the distance between both players is:

$$(Y-y)^2 + x^2 = L^2$$

and this never changes. You still head after him — it's what you have to do no matter how fruitless it will be — so you run at an angle where:

$$\tan A = \frac{y-Y}{x}$$

We know the winger's whereabouts — at least we know he'll always be in the centre of the pitch. And with that, the problem is solved. We know that $(y-Y)^2 + x^2 = L^2$, so the angle is:

$$\tan A = \frac{\sqrt{L^2 - x^2)}}{x}$$

Now set $\tan A = dy/dx$ to get:

$$\frac{dy}{dx} = \frac{\sqrt{L^2 - x^2)}}{x}$$

whose solution is the equation in the main text.

4 The probability that you get tackled in the *n*th step is *p*. This must have something to do with how far you go in that nth step, which we call *dx*. In other words, we assume $dp = dx/a$, where a – looking at the dimensions - is some length or distance. The total chance of making it n steps is *P(n)*, and we've said that $P(n+1) = (1-p)P(n)$. Making it *n* steps tackle free is the same as travelling a distance *x* without being tackled. If so, we can change this expression to $P(x+dx) = (1 - dx/a)P(x)$. Multiplying out, $P(x+dx) = P(x) - P(x)dx/a$. The difference $P(x+dx) - P(x)$ is the chance that you'll be tackled in the short distance between *x* and *x+dx*. This slight chance we call *dP*. Our equation is then $dP = -Pdx/a$ or $\frac{dP}{dx} = -P/a$. This has solution $P(x) = P(0)\exp(-x/a)$. As a final step, if you run an infinite distance someone, surely, has to tackle you. There is a 100% chance that you'll get dumped if you go infinitely far. In other words, by the time you reach infinity, *P=1*. This allows us to set:

$$1 = \int_0^\infty P(x)dx = P(0)\int_0^\infty \exp(-x/a)dx = P(0)a$$

This means that $P(0) = 1/a$ and so our probability is $P(x) = [\exp(-x/a)]/a$.
The average distance travelled before tackle is $\bar{x} = \int_0^\infty xP(x)dx$. Using our probability formula:

$$\bar{x} = \int_0^\infty xP(x)dx = \int_0^\infty x[\exp(-x/a)]/a dx = a\int_0^\infty ye^{-y}dy = a$$

So, the mysterious a in our equation is exactly the average distance you can expect to go before being tackled.

5 The exact solution is

$$T = \frac{2}{\sqrt{\left(\frac{u^2}{H^2}\right) + \frac{6g}{H}}}\left(K(k) - F(45^o, k)\right)$$

where *K* is the complete elliptic integral of the first kind; *F* is the incomplete elliptic integral of the first kind, and

$$k = \frac{1}{\sqrt{1 + \frac{u^2}{6gH}}}$$

For a full derivation, see the book by Bob Banks (*Towing Icebergs...*) in the list of suggested further reading.

6 Arguably the greatest person to blend rugby with physics was Sir Thomas Ranken Lyle. He began one season in the third XV of Trinity College, Dublin, and by the end of the same season was in the Irish national team. He played for them on four occasions, before suffering a knee injury that ended his career. But he was also a physicist, a pioneer of x rays, who was elected a Fellow of the Royal Society and knighted for his services to physics.

Chapter Four

Kicking, the habit: On penalties, conversions and Garryowens

'Tony Ward is the most important rugby player in Ireland. His legs are far more important to his country than even those of Marlene Dietrich were to the film industry. A little hairier, maybe, but a pair of absolute winners.'

C.M.H. Gibson

'I didn't know what was going on at the start in the swirling wind. The flags were all pointing in different directions and I thought the Irish had starched them just to fool us.'

Mike Watkins

There comes a time in any game when the ball has to be cleared. Usually it's obvious, and the boot of the outside half or a stalwart fullback does the job. Some fullbacks, such as England's Charlie Hodgson can send the ball sailing down the pitch more than 60 metres distant. When the pressure is on and you've been backed up in your own 22 for far too long, the hefty thump down the field into touch can provide much-needed relief.

In one high-school game our pack was hopelessly outgunned. We staved off five-metre scrum after five-metre scrum and desperately needed a break. To my surprise, I heeled one against the head. The scrum broke up and our pack looked up field to see how far away the lineout would be. Horror struck, for our fly half was running towards a welcoming committee. Two metres, if that, beyond our own try line their inside centre ripped the ball off him and trundled over for the score. Why no kick? His girlfriend was watching and he wanted to impress her. He expressed 'concerns' over the lack of support from the pack when he'd run into trouble and, as captain, I shared in what diplomats would describe as a 'frank and open discussion' expressing my wonderment that he didn't boot the ball away. It was a long afternoon…

But it's not just high-school teams who can ruin the kicking game. In the late seventies, the London team Richmond had a pretty good team. It's one of the oldest clubs in England, founded in 1861, and they played one of the earliest rugby games ever, against South-East London's Blackheath, which was formed a couple of years earlier in 1858. Richmond had a great sevens team, and often would do well in the Middlesex Sevens, even if they didn't win it. They even had a sevens specialist, Charlie Yeomans, who never truly made it into the 15-man game but was often the star of the sevens squad.

For Richmond, though, their number ten was the fly half in the ointment. For some reason, he had an aversion to passing. Time and again he'd thump the ball away, only to have the other team run it back. One of the Richmond wingers, Allan Mort, was one of the fastest creatures on two legs. On the rare occasions I saw Richmond play, courtesy of BBC2's Rugby Special, if Mort got the ball he was hard to stop. If only the fly half – my memory suppresses his name - had passed the ball down the line more often, or kicked the ball more strategically so that the wingers had something to chase, they might have been the best club to play in the Old Deer Park, and not play second fiddle to London Welsh.

HOW FAR CAN YOU GO: CLEARING THE BALL

The art of kicking is neatly tied to one of the oldest parts of physics: ballistics. Suppose you're in trouble inside your own twenty two. The task is to whack the ball as far down the field as possible, making sure that it will eventually trundle into touch. Imagine that the Biarritz and French fly half Damien Traille has to clear his goal line. What advice can physics dispense? The simplest model is that he kicks the ball with a speed v at an initial angle A to the ground. (See Figure 4.1) We want to work out what this angle should be for the ball to go as far as possible. To do this, use Newton's second law $\vec{F} = m\vec{a}$, where m is the mass of the ball which, according to the rules, is about 425 grams.

To make the calculations easy, neglect messy things like air resistance. In that case, in a horizontal direction travelling down the field, no force acts.

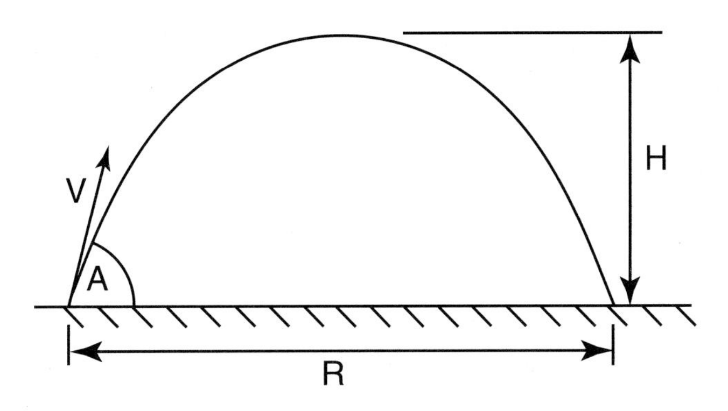

Figure 4.1: Projectile motion.

Vertically, the ball fights against gravity. If the ball is kicked with an initial speed v at an angle A to the ground, then the horizontal distance the ball travels before hitting the ground is its range, R. This is[1]:

$$R = \frac{v^2 \sin 2A}{g}$$

For the record, the maximum height the ball reaches is $H = \dfrac{v^2 \sin^2 A}{2g}$. If the ball goes straight up, then A is ninety degrees, which – no surprise – makes the height a maximum.

The biggest value that the sine function can take is 1, and this occurs when the angle is 90 degrees. So, Damien can kick the ball a maximum distance:

$$R = v^2 / g$$

which occurs when $2A = 90$ degrees. So, the best angle for him, or anyone else, to kick the ball for maximum distance is forty five degrees. When a top-quality kicker sends the ball some sixty metres down the field, then as g, the acceleration due to gravity, is roughly 10 ms^{-2}, the ball had a starting speed of around 25 m/s, a phenomenal speed close to 90 km/h or over 55 mph. This kick soars to a maximum vertical height of $v^2/4g$, about 15 metres and it hangs in the air for a time $T = (2v \sin A)/g$, which is about 3.5 seconds. So, a kick that clears your lines may go 60 metres or so downfield, but it will get there well ahead of your team mates, no matter how swiftly they run. That's why they use the quick lineout.

A phenomenal example of kicking for distance is Dan Biggar, one of the new wave of Welsh fly halves. Playing for the Ospreys against the Glasgow Warriors one wet weekday in September 2008, Dan got the ball on his own 10 metre line, and calmly proceeded to score a drop goal. Basically Biggar not only belted the ball at more than 25 m/s, but did so with enough accuracy to get the ball to go smack in the middle of the posts. Thanks to TV coverage by Sianel Pedwar Cymru, you can see the goal on YouTube.

For any aspiring outside half or fullback, the message from physics is loud and clear. Kick the ball at an angle of forty-five degrees. This you can practice by hanging a hula hoop from the branch of a nearby tree or lamp-post. Measure how high the centre of the hoop is off the ground. Then stand the same distance from the hoop and kick the ball. If it goes through the middle, you'll have been close to the optimal 45-degree launch angle.

The other lesson is also straightforward. The range depends on the *square* of the velocity. If you thump the ball so it goes *twice* as fast as normal, the ball will go *four times* as far. In a pressure situation, the key is to kick the ball as hard as you possibly can, rather than worry about launching the ball at the perfect angle. Suppose you aim poorly and, instead of kicking the ball at 45 degrees, it goes up at thirty degrees (i.e, you're off by 33%). If your maximum range is usually R, this time you'll kick it $R \sin 60$. The sine of sixty degrees, though, is $\sqrt{3}/2$, or about 0.866. So, for your poorly aimed kick, the distance is reduced only by about 15%. In contrast, if you completely flub the kick by kicking it at the perfect 45 degrees but with a launch speed only one third of your best, the distance is only $(4/9)R$, less than half your maximum range. The moral is, if you're going to kick at all, kick hard.

BOUNCING: BEYOND THE 22

One of the first rules of the game that neophytes learn - especially if they're in the backs - is that you can't kick the ball out on the full when you're beyond the 22 metre line. To make matters worse, the change in the laws says that you can't dart back inside your 22 just to thump the ball away. So, beyond the magic marker somewhere, somehow, the ball has to bounce. If your opponents happen to have inserted their fullback into the line and you strip the ball away, then a beautifully angled kick towards the tryline sets up a race between your backs and theirs. If you managed to kick the ball away quickly enough, then your winger has a good chance of outsprinting theirs to the ball, for the simple reason that your winger won't have to turn around before getting up a full head of steam. The bounce, though, can be critical.

Isaac Newton was more than just a pretty brain. He cared about practical things as well. He came up with the coefficient of restitution. On a perfect surface, the kind of thing that exists only in physics textbooks, a ball dropped from a height H bounces back to the same height H. In reality, as Isaac knew, this never happens. When the ball hits the ground, it deforms, ever so slightly, so some of its kinetic energy goes to restore the shape of the ball. Some of the energy produces noise, and yet more will go into heating up both ball and ground. All of these mechanisms, and more, drain kinetic energy from the ball and transfer it to the air or the ground. The simplest thing to do - which is the way physicists prefer to work - is to lump all these complicated effects together. This you can do with the coefficient of restitution, traditionally labelled by the letter e. If a ball hits the ground with a downwards velocity of v, it bounces back up with a speed ev. On its way down, the ball hits the ground with the kinetic energy $\frac{1}{2}mv^2$. It bounces up a moment later with speed ev, so it has lost a fraction $(1 - e^2)$ of its original kinetic energy. Alternatively, a fraction e^2 has been absorbed in the impact.

One of the fundamental laws of physics, that energy is conserved, means that if the ball was dropped from a height H, it bounces back to a height $e^2 H$. A perfectly hard surface has $e = 1$, so the ball bounces up to the height from which it was dropped. A thick, muddy surface, like Scotland's Murrayfield in the rain, has a coefficient of restitution closer to 0, so the ball hits the ground and sticks.

On any given day, conditions vary. The coefficient of restitution is not always going to be constant. This is the physics equivalent of saying beware of how the ball bounces. A ball landing with the same speed at the same angle on a bone-hard pitch will bounce higher than on a soft muddy one. If it rained in the morning and a stiff drying breeze swirls around the stadium but only part of the pitch is in sunshine, keep a sharp eye on the ball. The value of e is going to vary across the pitch, depending on how much water has drained into the ground or evaporated into the air. It could be a nightmare set of conditions for a fullback.

Ponder the restitution coefficient carefully. Suppose Damien kicks the ball at an angle A with a speed v. We know it will land at a distance R downfield, where $R = \frac{v^2 \sin 2A}{g}$. When it touches down, its horizontal velocity is $v \cos A$, which remains unchanged. Its vertical velocity, though, is $v \sin A$ on impact but this *does* change when the ball bounces. After bouncing, the ball travels upwards with a speed of magnitude $ev \sin A$.

Now solve the same equations as before, but for the new velocity. And then, when the ball touches down for a second time, it launches upwards with a new vertical speed, one diminished yet again by a factor of e. After N bounces, the ball has travelled a distance $R(N)$ downfield[2]:

$$R(N) = (1 + e + e^2 + \cdots + e^N)\left(\frac{v^2 \sin 2A}{g}\right)$$

where '...' instructs us to keep adding higher powers of e until we reach e^N.

There is something hidden, yet beautifully simple, in this equation. Suppose the ball bounces an infinite number of times. At first blush, it's tempting to think that the ball ends up an infinite distance from where it was kicked, in which case it's a drop out on your opponents 22. That's not the case. It so happens that multiplying by $(1 + e + e^2 + \ldots + e^N)$, as N heads towards infinity is precisely the same as *dividing* by $(1 - e)$. So, after an infinite number of bounces, the ball has gone a total distance R, where

$$R = \frac{v^2 \sin 2A}{g(1-e)}$$

Another way to view this is to use D, the distance travelled by the ball before it bounces for the first time. In that case, the total range is $R = D/(1-e)$. On a perfectly hard pitch, e is 1. Kick on that surface and a ball that bounces after a mere five metres will continue bouncing beyond their try line and generate a 22 m drop out. So, when playing in the Dubai or Hong Kong Sevens, in the blistering heat on a hard dry pitch, the ball will bounce a long way – ease up on your kicking speed! In the middle of a downpour at Maesteg's home ground, e will be far closer to zero. A dry pitch has a higher e than a wet pitch; long grass will cause a lower e than short-cropped grass, for the ball has to brush aside the blades of grass before rebounding. A quick experiment with a metre ruler, a rugby ball, and some of my children suggests that on regular grass, the coefficient of restitution is about 0.6.

If you're a fullback whose job is to hang back waiting for the long kick, or a back whose job it is to supply the long kick, you might want to eyeball e before the game to get a feel for how far the ball is going to bounce, all other things being equal. To do that, hold the ball over your head and drop it onto the ground, making sure it doesn't bounce on the pointy bits. Suppose the ball bounces back up to your knees. Remember that the ratio of the height from which you dropped the ball to the height it rebounded to is e^2. Your knees are about a quarter of the distance to the tips of your outstretched fingers, which would make e about 0.5. So, if you can kick the ball a distance D on this surface before it bounces, it should carry about $2D$ before coming to a halt.

Rugby is never that simple. If we played with a round piece of pigskin, like soccer players do, all would be well. A prolate spheroid, though, is a nasty piece of work when it bounces. Landing amidships makes it bounce differently from when it touches down on a point. As the ball bounces awkwardly over your head and rolls behind you, try to think of it as one of the joys of the game, even if your opponents have just scored.

The way the ball bounces

Rugby is certainly played with an odd-shaped ball. There's nothing odd about the shape, though – not really. It's closely related to one of the great curves of geometry and physics, the ellipse. For centuries humans thought that the Sun moved around the Earth. Then Copernicus suggested, rightly, that the Earth moves around the Sun. The problem was that scientists still believed, in a certain sense, that the heavens were perfect and so tried to reconcile this Sun-centreed view with a circular Earth orbit — the circle being the perfect shape with no beginning or end. It was Johannes Kepler, whose mother was suspected of witchcraft, who proposed that planets might move in an ellipse.

The instructions on drawing an ellipse are simple. Suppose you have a point *A* and a point *B* that you've marked on a sheet of paper. Another point, *P*, is on the ellipse if the distance from *A* to *P* (*AP*) added to the distance from *B* to *P* (*BP*) is a constant. This is the kind of phrase that only a maths teacher could love. But if you feel like drawing a giant rugby ball on the middle of the pitch in the middle of the night, it's not too hard to do. Two pegs, a rope, a brush, and some white liquid suffice. Bang the two pegs in the ground (*A* and *B*). Get a rope. Tie one end to peg *A*, the other end to peg *B*. Get a paint brush (*P*) or whatever you're going to draw the ball with. Place the brush against the inside of the rope and pull it taut so that the two pegs and your hand form a triangle. Now move your hand, keeping the rope taut. The result: a beautiful rugby ball shape in the centre of the field —and a livid groundskeeper.

In real life, a ball is not a flat ellipse. It's as though someone's taken an ellipse and rotated it about the long axis. This three-dimensional object is an ellipsoid or, more precisely, an oblate spheroid. Mathematicians know that if the end-to-end length of the ball is $2a$ and the maximum radius of the ball is b, then the volume of the ellipse[3] is $(4\pi/3)ab^2$. When you play soccer, the ball is spherical and so $a=b$, which means the ball's volume is $(4\pi/3)b^3$, as it should be.

To make a rugby ball, you need to have a certain amount of leather handy, which forms the surface area of the ball. Strange to say, while the volume is fairly straight forward, the surface area is a mess. The expression for the area involves such things as complete and incomplete 'elliptic integrals of the first kind', which many people with Ph.D.s in physics know nothing about. (And yes, sci. fi. fans, there are elliptic integrals of the third kind!). We can make a reasonable estimate. The surface area has to be less than that of a sphere whose diameter is the end-to-end length of a rugby ball. By the same token, it has to be more than a sphere whose radius equals that of the middle of a rugby ball. The area of a sphere of radius R is $A = 4\pi R^2$, so we know that $4\pi b^2 < A < 4\pi a^2$.

Balls nowadays are possibly made with leather, but often of synthetic materials, like polyurethane. The International Rugby Board specifies the ball quite precisely, above and beyond specification of the material. The 'length in line', which is $2a$ in our equation, must lurk between 280 and 300 mm. The circumference is between 580 and 620 mm, which is $2\pi b$. These suggest that a = 15 cm and b = 9.8 cm. This means that the volume of the ball is about 6,000 cubic centimetres.

The area we've guessed as being in the range 1200 cm^2 < A < 2827 cm^2. This is quite a broad range, but we can do better. Mathematicians show us that a 'good enough' approximation, one that doesn't involve elliptic integrals, is:

$$A = 4\pi\left(\frac{2a^p b^p + b^{2p}}{3}\right)^{1/p}$$

where, if p=1.6, we're within 1.2% of the exact answer[4]. Our 'good enough' answer when p =1.6 is A~1600 cm^2, which you could use to cover a cube of side 16 cm.

Over the years, the ball has changed. William Gilbert, who died in 1877, used to make balls for the Rugby School. The original ball required a fresh pig's bladder, ensconced in leather, which some poor soul had to blow up by use of a clay pipe. His business rival, Richard Lindon, managed to persuade Mrs. Lindon to perform this task. She eventually died of a lung disease, possibly transmitted through the pipe from an infected pig's bladder. Perhaps in the grip of remorse, her widower invented the artificial pig's bladder and a brass hand pump, which got rid of the need for the clay pipe. No doubt still beside himself with grief, he failed to patent either invention, missing out on a pile of cash but letting the world inflate their rugby balls free of virulent pig-inherited illness.

The IRB lays down the law regarding air pressure inside the ball. Whether you pump it up by mouth, clay pipe, or brass hand pump, it has to be between 65.71 and 68.75 kiloPascals above atmospheric pressure (otherwise known as the gauge pressure). This pressure can change during the game. If the ball is pumped up to 68.75 kPa in a warm equipment room (the ref roasts at 300 Kelvin or 27 Centigrade) and is then brought down to the freezing field at 273 Kelvin (0 Centigrade), strange things may happen. If air is roughly an ideal gas, which it is, and the volume of the ball doesn't change (the leather or synthetic skin won't let that happen), then we can use the ideal gas law for the pressure P, volume V, and temperature T, which is $PV/T = const$. As the volume is constant, we have $P/T = const$, a.k.a. Gay-Lussac's law, discovered in 1802 by the French scientist Joseph Louis Gay-Lussac. If the air in the ball began life at 300 Kelvin and, by the end of the game has dropped to match its surroundings, then the pressure will decrease by about 9%. In other words, the ball that seemed so hard at the beginning of the game may feel flat towards the end, for the gas inside exerts a pressure of only 52.5 kPa, well below the legal limit.

The problem with an underinflated ball is that it's hard to kick. As you apply boot to ball, the softer ball can deform its shape more easily, so more of your kicking energy goes into temporarily deforming the ball and less into making the ball move. In the RWC 07, England and Australia's quarter final had the opposite problem: over inflated balls. England's Wilkinson and Australia's Stirling Mortlock missed three penalties apiece in the clash that saw England move forward to the finals.

As far as understanding how the ball bounces, we'll never have much luck. Physicists have argued in print about how a stiff rod bounces off of a completely flat and infinitely hard surface. A rod is easy, though. All you have to worry about is the vertical velocity and the rotational speed when one of the points hits the surface. For a rugby ball, it depends on both of those things, it's true, but also on which part of the ball touches the ground which, of course, is not a hard flat surface. Simply put, there's no way the simple laws of physics can deal with the motion of something as complex as a rugby ball. Suffice it to say, the motion is chaotic. Change the bounce ever so slightly – either velocity, rotational velocity, or the patch of leather on which the ball lands – and the outcome will be vastly different. A centimetre difference can mean the ball will sit up nicely off the turf so that your full back can gather it, or scoot by him beyond the try line, where he's beaten to the ball.

The best that physics can do is to offer a simple observation. If you roll the ball off the end of your boot, so that sometimes you can see the word 'Gilbert' as you look at it, it has a good chance of rolling sedately across the perfectly manicured turf. If you want it to move unpredictably, then drop it onto your boot with the word 'Gilbert' vertical. Then as it rolls along the long side, it can suddenly kick up quite nicely for your winger to take in full stride.

That said, even the ball rolling quietly and peacefully on a flat surface is difficult to cope with mathematically. To describe the path, you need to know the perimeter of an ellipse which, like the area of an ellipsoid, is a mathematical nightmare. (Still, many people enjoy studying such problems, simply for the intellectual challenges they provide. Mathematicians have made a bike with square wheels that rides smoothly over a surface that is shaped like an inverse catenary – a catenary being the shape of a hanging chain.). In this age of computers, you can find a lovely animation on the web of an ellipse rolling over a flat surface – the kind where in a game you wonder nervously if the ball is going to keep on rolling into touch, or into the in-goal area, or stop short. Enjoy!

But let's not forget the mathematicians. Charmed by geometry as they are, they can tell your groundskeeper how to prepare a pitch so that a rolling rugby ball will always have its midpoint the same height above the surface. In 1872, Arthur George Greenhill, a mathematician at the Artillery College in Woolwich solved the problem. If the groundskeeper can shape the pitch so that the surface is the 'delta amplitude of the Jacobian elliptic function', life would be far easier! Sad to say, the workers at the Woolwich Arsenal preferred Association to Rugby football. When the team relocated to North London, they dropped 'Woolwich' from their name and have, so football fans tell me, done quite well. I was born in Woolwich, where we keep the memory of the team alive and refer to them as '*The* Arsenal.'

Another mathematician, Cambridge's Keith Moffatt, teamed up with Japan's Yutaka Shimomura to explain how a hard-boiled egg can begin spinning horizontally but end up spinning vertically instead. The two scientists treated the egg as an ellipsoid, which is a far better model for a rugby ball than an egg (which is more akin to the bottom half of a sphere stuck on to the top half of an ellipsoid). Try spinning the rugby ball on the clubhouse floor and see what happens. After all, it's such an important piece of science that Professors Moffatt and Shimomura had their work published in the journal *Nature*, arguably the most prestigious science publication in the world.

Oxford University played a crucial role in the early years of rugby. William Webb Ellis attended Brasenose College and OU RFC was founded in 1869. They beat Cambridge 1-0 in the first game between the two universities (i.e., by a converted try to nil). In 1936, the only bright spot of the New Zealand tour of the British Isles was afforded when the Oxford student and white Russian noble Prince Alexander Obolensky scored two tries in England's 13-0 victory. Cambridge, though, has won more of the annual battles than the Dark Blues, particularly in recent years. Could it be that Professor Moffatt's research on the motion of ellipsoids has resulted in some important results that he has shared not with the world of physics but only with the Light Blues First XV?

THE HABIT OF KICKING

A healthy wallop of the ball takes you four times farther down the pitch than a tentative timid tap. Then, if the luck of the bounce is with you and the ground is hard, the ball will keep on rolling. The question, though, is how to increase the speed at which you thump the ball. Clearly, to go long you need to enhance the launch speed. Again, physics can construct a fairly simple description of what's going on.

Suppose England has scored. Wilkinson makes a lovely little mud pie, or uses one of those plastic thingies. He's taken the famous three steps back and two sideways. To look the part, he throws a few blades of grass into the air to judge the wind speed. And then he starts to run.

Take a snapshot as boot reaches ball. (See Figure 4.2) Our stalwart kicker has one foot firmly planted on Planet Earth. The other one is swinging on the end of a leg that rotates swiftly around the hip bone. We can now break out conservation of momentum – but with a twist. We assume here that *angular* momentum is conserved, when added up around the hip bone. To do this, the hip bone has to be stationary (i.e., it rotates but does not otherwise move). This means instead of Jonny running with speed u to kick a stationary ball, we imagine a stationary Wilkinson kicking a ball that comes to him at speed $-u$. If the England fly half as a leg of length L, the angular momentum of the ball about his hip, prior to kicking, is $-mLu$. The angular momentum of his leg prior to the kick is $I\omega$, where I is the moment of inertia of the Wilkinsonian leg and ω (the Greek letter omega) is its angular frequency. A moment later, his leg rotates with angular frequency Ω (the Greek capital letter omega) and the ball shoots off with angular momentum mLv. Model his shapely leg and well turned ankle as a rod of length L and mass M so that $I = ML^2/3$. As with the stiff-armed pass, we save trees by defining $x = 3m/M$. Then say that, when boot hits ball, only a fraction e of the energy goes into the motion of the ball

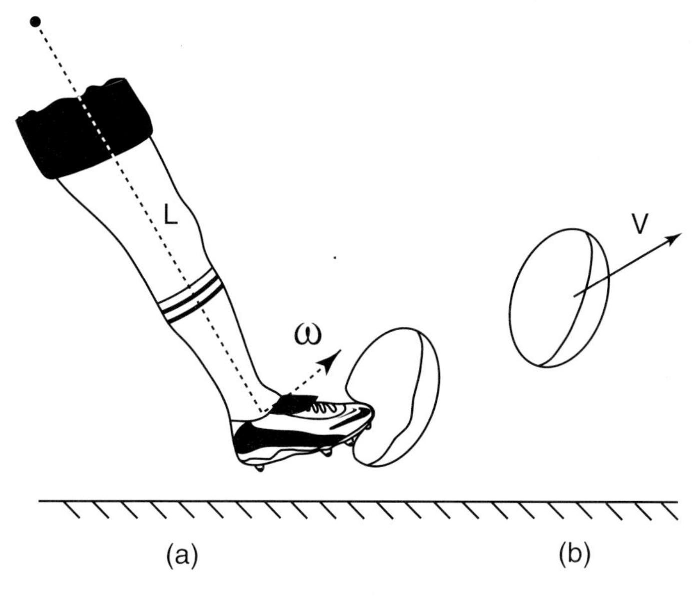

Figure 4.2: Kicking motion: The ball compresses on kicking.

Our rough and ready estimate is that the ball leaves Jonny's boot with a speed[5]:

$$v = u + e\left[\frac{2\omega L}{(1+x)} + u\left(\frac{1-x}{1+x}\right)\right]$$

The formula here looks good. First, the faster you run up to kick the ball, the bigger u will be and the higher the ball speed. Second, the swifter you rotate your leg – the faster you can kick the ball – the greater the angular frequency

ω and so the farther the ball will travel. Last, the longer your leg, the faster the ball will go. In other words, it pays to stretch that leg out when you kick the ball. Also, when $e = 0$, the ball sticks to the kicker's foot. In that case, there is no kick, and the only speed the ball will have is the speed at which Wilkinson runs.

Doctors assure us, a la Frankenstein, that a human leg is about 10% of a person's mass. If so, then many a rugby team has been ruined by the 8.8kg right leg that belongs to England's fly half. According to law 2, the ball has a maximum mass of 0.440 kg, so that $x = 3m/M \sim 0.15$. Video tapes suggest that kickers make contact when their leg is rotating about 20 radians/second. Last, sports scientists have analysed the deformation of a rugby ball when kicked, and think its e value is about 0.5.

If Scotland harasses Wilkinson so that he has to clear inside his own 22, and in a hurry, he won't have time to run. In that case, $u = 0$ and the launch speed is about 17 m/s, which is 61 km/h or about 38 mph. This would dispatch the ball about 30 metres downfield. On a conversion attempt, assuming he can get up to 5 m/s before kicking the ball, the launch speed is 24.3 m/s, increasing the range to 59 metres. The results, again, should not be taken too literally. If we change e to 0.6, then the ball kicked while stationary now goes about 44 metres while the ball kicked while at 5 m/s tilt now goes 79 metres. The details are not too important. We've identified, though, that kicking is a process in which angular momentum is conserved; the results are plausible, and so physics then dictates what we need to do to make the most of it.

And finally, some torque talk. As you move your right foot back to kick the ball, you become unbalanced. Your weight of your leg creates a torque that tries to twist your torso. To kick the ball farther, you need to be as well balanced as possible. The remedy: a countertorque. As your right leg goes back, throw your left arm forwards and upwards, just like the professionals.

ALL HIGH AND MIGHTY

Thumping the ball hard downfield can be immensely satisfying. Launching a bullet that pushes your opponents back some 60 metres relieves pressure on your team and, in the dying moments of the game, eats up valuable time on the clock. Sometimes, though, maximum distance is not what's called for. A case in point is a kick off or restart. Another is a crossfield kick from, say, your winger. In both cases, you want the ball to go a certain distance, but in its own good time. You want the ball to land as far down the pitch as possible, while allowing your colleagues time to assemble either to snag the ball or put extreme pressure on the luckless opponent who has to gather it in. As usual, physics can explain the game.

The ball travels a distance:

$$R = \frac{v^2 \sin 2A}{g}$$

This is the same formula that governs the clearance kick downfield. Suppose one of their players has coughed up the ball on a previous restart. We want to aim for him again and so need the kick to go 10 metres downfield and, say, roughly 18 metres over (halfway between the middle of the pitch and the touchline). This is the point, X marks the spot, where we wish our pack to assemble just as the nervous shaky catcher attempts to gather the ball. Courtesy of Pythagoras R is set at about 20 metres. Your fly half kicks the ball at, say, 20 metres per second. According to the equation, the number 10 has to kick the ball at an angle A where $\sin 2A = 1/2$. At first glance, this seems easy. After all, sin 30 equals 0.5, so the fly half has to kick at 15 degrees. What is not so obvious, though, is that he has a choice, for sin 150 = ½ as well[6].

What this shows is that for a given distance and a given speed of kicking, there are two angles the fly half can kick the ball at. If one of these is A, the other is 90-A. The two choices have different outcomes. The ball travels the distance R in a time T, which is predicted from the formula:

$$R = (v \cos A)T$$

He can welly the ball at an angle A or at 90-A. It'll get to exactly the same position, but at different times. If A is less than 45 degrees it's a low kick, and the other angle will be more than 45 degrees – be inventive and call it the high kick! As R and v are the same for both kicks, the times are governed by:

$$\frac{T(high)}{T(low)} = \frac{\cos A}{\cos(90 - A)}$$

A mathematical fact of life is that cos (90-*A*) = sin *A* so:

$$\frac{T(high)}{T(low)} = \cot A$$

In our example, the high kick goes up at an angle of 75 degrees, the low kick at 15 degrees. As cot 15 = 3.733, the high kick stays in the air between three and four times as long as the low kick. So, if you want to give your team time to gather under the ball as it lands – kick high! The other advantage is that such a kick will get you over large obstacles – such as second row forwards. These two kicks go far different heights. (See Figure 4.3) The ratio of the maximum heights is

$$\frac{H(high)}{H(low)} = \frac{\sin^2(90 - A)}{\sin^2 A} = \frac{\cos^2 A}{\sin^2 A} = \cot^2 A$$

For our simple example when *A*=15 degrees, the maximum heights differ by a factor of about 14. It comes as no surprise, then, that to put pressure on the opposition as they catch the ball, kick it high and hard.

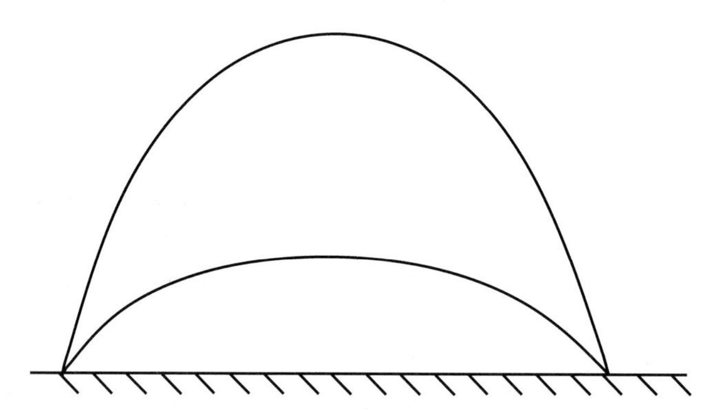

Figure 4.3: Two kicks: Same speed and range, different heights and hang times.

Summed up, if Wilkinson kicked the ball high from a restart, he could let Jason 'the Fun Bus' Leonard run four times as slowly as a shallow kick and still get to the ball. On the other hand, imagine having to keep your eye on a ball that goes about 20 metres high and keep concentrating for almost 4 seconds while the ample 111kg of England's most capped player heads towards you at flank speed.

Don't forget that this is not just for restarts. In the 2007 Rugby World Cup, when England played Tonga, England had a penalty that Wilkinson was surely going to kick between the posts. On the far side of the field, though, he saw that Phil Sackey's marker had drifted away. A lovely lofted kick from the English number 10 went the width of the pitch, into the in-goal area, where Sackey had enough time to catch it and fall on the ball before rolling out of bounds. A truly lovely try.

The high kick, the Garryowen (named after the County Limerick club for whom Tony Ward and, in his younger years, actor Richard 'Dumbledore' Harris both played) can serve another useful purpose. When the Worcester Warriors went up against the heavily favoured London Wasps in September 2008, a superb flighted kick from Warriors fly half Matthew Jones caused an upset. England's Josh Lewsey lost the ball in the late summer sun and fumbled the ball backwards, a mere 10 yards from his own line. The Worcester centre Sam Tuitupou was the swiftest to react, earning the Warriors five points. They would go on to earn a surprise victory over the Premiership's reigning champions.

Full backs in drag: Air resistance and the Magnus effect

If Mr. Leonard gets to clear the ball, though, he'll probably do what all props do: hold the ball with the laces and letters facing him. He'll then punt the ball down field, making sure that he kicks it with the top part of his boot. A real kicker, like Italy's (or Argentina's) Diego Dominguez, usually strikes the ball with the laces and letters pointing away from him. He'll make contact using the side of his foot, so that the ball is launched with a lovely spiral motion. This serves a useful purpose; it helps the ball go where the kicker aims it.

The spinning top is a favourite children's toy. It spins about an axis. If you tap the top, then an odd thing happens: it wants to stay spinning in exactly the same direction as it started spinning. In the language of physics, if you impart a torque about a certain axis, then the object wants to remain rotating about that axis. This phenomenon is known as gyroscopic stability, and as its name suggests, it is the fundamental principle of the gyroscope. In the eighteenth century, Benjamin Robins (the first engineer general of the East India Company) realised that this could be put to good military use. Muskets were not particularly accurate; by cutting a spiral down the inside of the barrel, a bullet could be given a spinning motion as it left the muzzle. This enabled the bullet to go where it was aimed. The process by which the spiral was cut into the barrel was known as rifling, and the guns that were produced were, naturally enough, called rifles.

Gyroscopes are what you use in outer space in order to sense direction. There's no convenient way to determine up and down in the weightlessness of space, but a gyroscope will always want to point in the same direction. This is what brought the astronauts of Apollo 13 safely home. It's this same cutting edge, space-age technology that a fly half uses when kicking the ball. The spin Diego imparted to the ball gave it a wonderful gyroscopic stability, and it continues to go in the direction in which it was originally kicked. The prop's effort, punted crossways, won't have such stability and may wobble unpredictably through the air, not going where he wanted it to go nor anywhere near as far.

The great advantage of having a ball that spins is that it wants to stay aimed at the angle at which it was launched. So, if the ball was kicked at an angle of 45 degrees relative to the ground, with its long axis pointing at an angle of sixty degrees to the ground, it wants, as a consequence of the gyroscopic effect, to stay at an angle of 60 degrees at any point along its path. (See Figure 4.4a). In the language of aerodynamics, the angle between the ball's direction and the angle of the ball is the angle of attack or, for pilots, it's yaw. When you have an angle of attack, a lift force acts on the ball. This lift force counteracts the pull of gravity, so the ball sails farther through the air before touching down. Compared with the gravitational force, lift is more significant at high speed, in low-altitude stadiums, and dry climates. With a lift force to counteract gravity, the ball will travel much farther if it is kicked off the side of the boot. (That said, there's a corresponding drag force to slow the ball down and which acts parallel to, but against, the ball's forward motion).

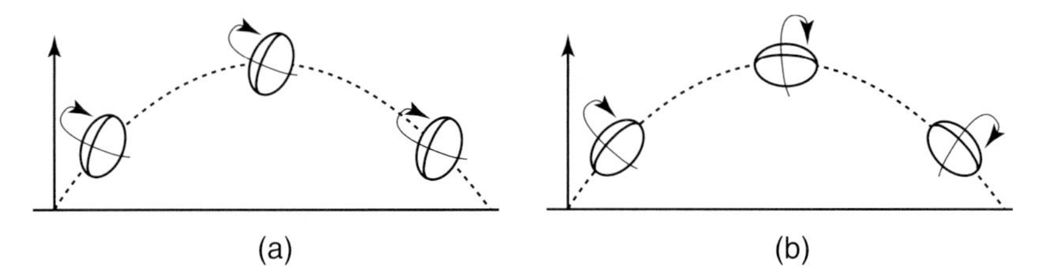

Figure 4.4: A gyroscopic kick (a) and a torpedo punt (b).

The size of the drag force depends on the square of the ball's speed, the ball's area, the ball's shape, and the density of the air. Specifically, $F = \frac{1}{2} CA\rho v^2$.

Here C is the drag coefficient, which depends on the shape of the ball, A is the surface area, and the Greek letter rho, ρ, is the density of the air. Double the ball speed, you quadruple the drag force. The larger the area pointing into the wind, A, the larger the drag will be. Last, the higher the density of the air, the bigger the drag force is. So, drag is more of a problem on humid days when you're playing at low altitude. (See Figure 4.5)

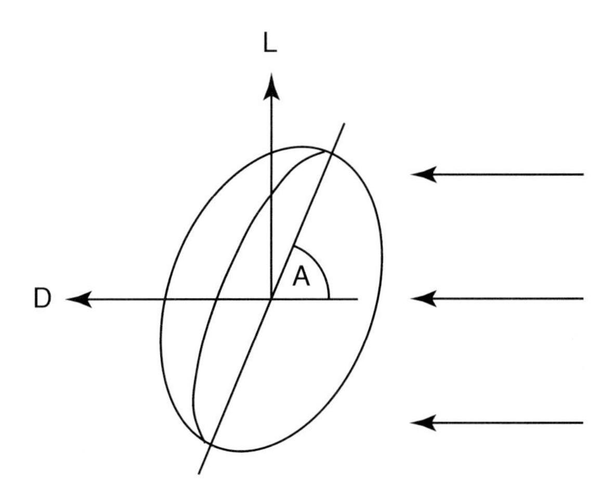

Figure 4.5: Lift (L) and drag (D) on a ball that has an 'attack angle' A.

When Gavin Hastings (who has the record for the most conversions in the Rugby World Cup) played for Scotland, the full back often had to clear his lines by kicking far down the pitch. He had no choice about v, he needed to make it as large as possible. Mr. Hastings can't change the shape of the ball, nor the temperature, humidity, and thus the density of the air. The area A, though, depends on how he kicks the ball. The rugby ball, technically speaking, is a prolate spheroid. A prop, who punts the ball amidships, makes the area A almost that of a rectangle, whose length is the point-to-point length of the ball and whose width is the diameter of the ball at its fattest point. For the ball I have at home, these are 29.1 and 7.48 cm respectively. This is an area of 217.7 cm^2. If a back kicks the ball and makes a complete mess of it, so that the ball rotates end over end, then this, too, will be its surface area. The only difference between the two is that the completely messed up kick should be tougher for your opponents to catch cleanly.

Kicked in a tight spiral, the ball looks – to the air – like a circle whose diameter is the same 7.48 cm. [See Fig. 4.4(b)]. This area is only 43.94 cm^2. In other words, the drag force is reduced by a factor 217.7/43.94 – by about 5 times.

There is another consequence of gyroscopic stability, the spinning motion with which the ball moves. Formally, it's known as the Magnus effect, and it's what happens when velocity meets rotation. (See Figure 4.6) Suppose the ball whizzes at velocity $\vec{v}$ and rotates with an angular velocity $\vec{\omega}$. If we think of a rugby ball as a cylinder of length L, radius R, travelling through a gas of density ρ, then the Magnus force[7] on the ball is roughly $F = \pi \rho v R^2 \omega L$. For football fans, it's the Magnus force that lets you bend it like Beckham; for Australians, it helps explain the motion of a boomerang; and for Americans, it's a motivational force in Frisbee throwing. The direction of the Magnus force, for those who know their vector products, is $\vec{\omega} \times \vec{v}$. This indicates that for a torpedo punt (Figure 4.4b), where the rotation is always parallel to the velocity, there is no Magnus effect at all.

The more savagely the ball is kicked, the greater v, and the more the ball will swerve. The more rapidly the ball spins, the more it bends. Likewise, the Magnus effect is more pronounced in sea-level stadiums; at high altitudes, the density of the air will be lower (such as when the USA rugby team bested Uruguay 43-9 at the Rio Tinto Stadium in Salt Lake City, some 1,288 metres above sea level).

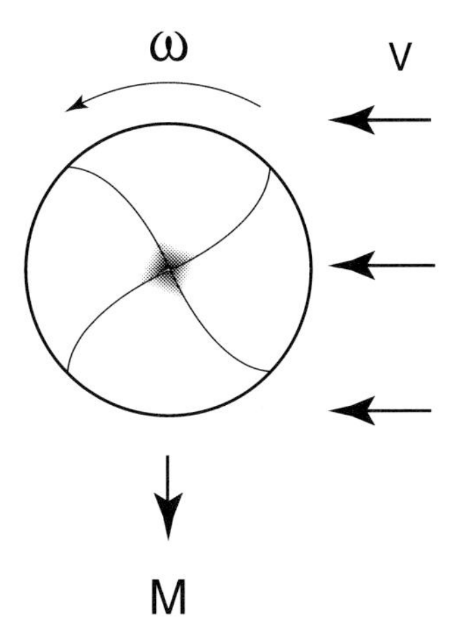

Figure 4.6: A spinning ball in a cross wind has a Magnus force, M.

There's a problem for the full back who has to catch the ball, for he has to remember who kicked it. A left footer such as Andy Farrell (the former England rugby league captain who changed not only codes but positions, deserting the League forwards for the Union backs) kicks the ball with a clockwise spin (as he sees it). South Africa's Andrew David 'Butch' James will send the ball rocketing away with an anti-clockwise spin. This means that their balls may have the same rotation rate, but a different sense and so the Magnus force acts in opposite directions. And if that wasn't bad enough, a crosswind contributes a Magnus force of its own, in a direction that either enhances or detracts from that of gravity. If the wind blows from the right as you see it, then Andy Farrell's ball has a Magnus force that acts upwards, letting it float farther downfield, while Butch's ball will dip down sharply.

As good physicists, we should check the size of the Magnus effect. Sports science research shows you can kick or throw a rugby ball with a value of ω of about 5 revolutions per second, which is about 30 rads/sec. If we're playing Italy in the Stadio Flaminio in Rome, where the average wind speed is about 5 m/s, then the ratio of the Magnus force to gravity is:

$$\frac{F_{MAGNUS}}{F_{GRAVITY}} \quad \frac{\pi\rho_{AIR}R^2 Lv\omega}{Mg} \sim \frac{3\times1.2\times(0.1)^2\times(0.15)\times5\times30}{0.425\times10} \sim 0.2$$

While we shouldn't take these numbers too literally, they suggest that, with its drastic dips and sinister swerves, the Magnus is a force to be reckoned with. Bear that in mind when kicking, or catching, on windy day.

A SEVERE CASE OF WIND

Rugby doesn't stop just because there's a breeze blowing. Ask the home team who play in Westpac Stadium, New Zealand, dubbed the windiest rugby ground in the world – they're called the Wellington Hurricanes! Rumour has it that the French managed to kick a wind-assisted drop goal from a phenomenal 87 metres out against the home side. Strange to say, a strong wind can wreck your chances of scoring points, but it cannot wreck the equations of physics. Suppose you kick the ball, as before, with speed v at an angle A to the ground. There's a breeze blowing of u m/s pointing straight towards you as you kick. Vertically, the equations we used before are unscathed. Horizontally, things are a tad different. Instead of tracing out a distance x in time t, where

$$x = (v\cos A)t$$

there's a new distance covered. The new horizontal velocity is the ball's speed minus the wind speed, so that

$$x = (v\cos A - u)t$$

The combined effect is that the ball now has a maximum range of

$$R = \frac{v^2 \sin 2A}{g}(1 - \frac{u}{v\cos A})$$

Calculus lets us work out the optimal launch angle. It is[8]:

$$\cos A = \frac{1}{\sqrt{2}}\left(1 + \frac{u^2}{8v^2}\right) + \frac{u}{4v}$$

Again, physics and rugby coincide. When there's no breeze blowing, $u = 0$ and the optimal launch angle and maximum range is exactly the same as before. If the wind blows stiffly, at about 25% of your normal kicking speed, the best launch angle is about 39 degrees. You'll then be able to kick the ball about 2/3 the distance of your wind-free best. (See Figure 4.7). In the second half, with the wind at your back, the situation changes. You ought then to thump the ball at roughly 49 degrees, and it will journey about 38% farther than in windless conditions.

The highest wind speed ever recorded in the United Kingdom (at low level, not in the mountains) was in Aberdeenshire and the anemometer recorded 228 km/h. This is 228 x 1000/3600 = 63 m/s. If you kick the ball at 21 m/s under wind free circumstances, it would travel a maximum of 44.1 m. The same kick on that fateful February 13, 1989 in Fraserburgh would have gone at most 30 metres downfield against the wind. In the second half, the same kick could go 65 metres. Ignoring the details, though, the big picture suggests that you aim lower when kicking into the wind, and higher when the wind's at your back.

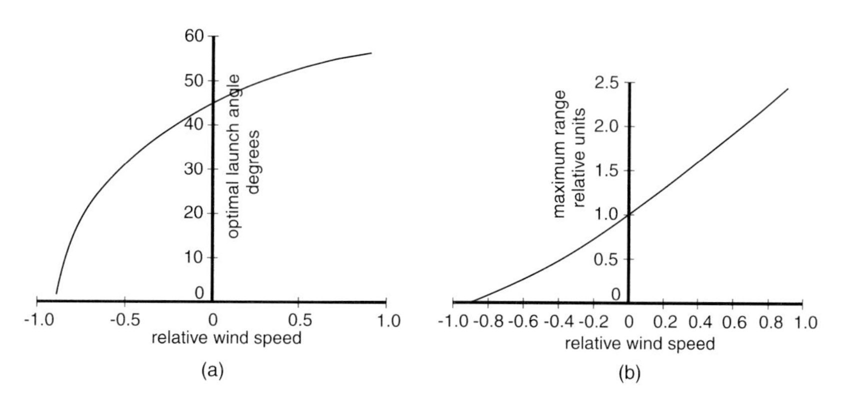

Figure 4.7: (a) Optimal launch angle versus relative wind speed. (b) Maximum range versus relative wind speed.

This is a nice model, but requires a good strong healthy and constant wind. It's the kind of conditions you get playing on a pitch at the local park. It's more complex if you're in a stadium. Here the wind can batter against the stadium wall and a strange event occurs. On the inside of the wall, whirlpools of air are formed. Slowly they grow stronger and stronger until, unable to stand it any more, they leave the wall and sweep across the pitch. This is known in the fluid dynamics trade as vortex shedding. For kickers, it's a nightmare. It means that a strong gusting winds appear from nowhere, usually as the ball heads towards the posts, and then mysteriously die away. All will be relatively calm until the next vortex has worked up enough strength to leave the wall. (See Figure 4.8).

Figure 4.8: A von Karman vortex sheet. Giant whirlpools of air (diameter roughly the size of the stadium wall) make for gusty kicking conditions.

You can see this to great effect on the flags above the stadium. The fluttering of flags occurs because the wind flows around the flagpole, vortices build up, and are shed. This forms a little alley – called the von Karman vortex street – of whirlpools downwind of the pole. The flag moves in between the eddies. Each time a vortex is shed, the pole vibrates a bit and the flag flaps, which is why flag flapping is one of the marks on the Beaufort Scale, used to assess wind speed.

Kickers aren't the only ones who have to worry. Just as the flagpole vibrates, so too does the stadium. The frequency f at which vortices are shed is governed by the Strouhal number S. If a stadium has diameter D and the wind speed is v, then the vortices are shed at the frequency $f = vS / D$. S is a number, roughly 0.2. Suppose the wind is a stiff 60 mph, or 27 m/s. The Millennium Stadium has a footprint of about 40,000 square metres. If we assume it's circular, which it is isn't, the diameter is $2\sqrt{40,000/\pi}$, or 225 metres. The frequency is then 0.02 Hz. This means a vortex is shed every 300 seconds or so. If the frequency happens to be the frequency at which the stadium vibrates, the building can collapse. Architects plan accordingly, to avoid this. We can sit safely in our seats to see the game but, for the players, there's a more practical concern. Every 300 seconds, about 5 minutes, a giant swirl of air will be kicked loose and sweep across the stadium, wreaking havoc if you've just kicked the ball towards the goal.

Putting three points on the board – or converting a try

My brother, born in Wales, is fond of telling me that Wales beat England in the 2003 World Cup; it was just that England scored more points. Any rugby player or rugby fan has had similar experiences. Your team has scored more tries, played expansive, entertaining, open rugby, only to lose to a team whose kicker had a boot as deadly as Jonny Wilkinson's. Wilkinson, who heads the tables for the most penalties and drop goals in RWC tournaments, can single-footedly destroy opponents. The truth be told, penalty kicks are a key part of the modern game of rugby, even though we do prefer to see points scored by wingers diving over for a last-minute score in the corner.

Penalty kicks, or conversions, are more challenging than trying to clear your lines. In a pinch, even a prop could probably get sufficient energy into a kick to muscle the ball a fair distance downfield. You probably wouldn't want to have that same prop take a penalty. (When playing tight-head prop in high school, I did get to attempt a penalty kick from just beyond the twenty two. It's amazing how far apart the two posts are until you put the ball on the ground. Then when you back up to kick it, the distance between the posts shrinks to nothing. I managed to get the three points, but it cured me of wanting ever to try kicking again. How Wilkinson, Hook, and the others do it so reliably and confidently, I'll never know).

With a physics-eye view, there is a distinct difficulty in place kicks. Not only does the ball have to travel a certain distance to the try line, it has to cross the try line at a certain minimum height - at which the crossbar sits. This is more complicated than the simple range equation that we've used up to now, but it is still solved fairly cleanly[9]. We know that if the maximum distance you can kick a ball is R , then the maximum distance you can kick from and score is L, where:

$$L = R\sqrt{1 - \frac{2H}{R}} \approx R - H$$

In 2005, Gavin Henson broke English hearts. With little time left on the clock, his monster penalty kick sailed over the bar and between the posts to gain a memorable victory for the Red shirts. Suppose Gavin normally kicks the ball some 60 metres before it touches the ground, so that $R = 60$ m. According to regulations, the crossbars are at a height H of 3 metres. So, putting in the numbers, Gavin can score from 56.9 metres away. (See Figure 4.9). (The honours here go not to Henson but to another Welsh player, Paul Thorburn. While playing against Scotland, he thumped a penalty over from some 62 metres out which, to use the phrase of the anguished Scots commentator Bill McLaren, was 'a monster kick.')

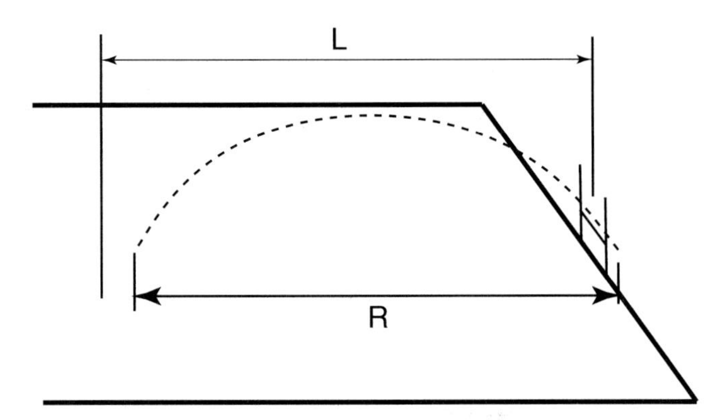

Figure 4.9: If your maximum range is R, you can score from L ≈ R-3 metres away.

The numbers illustrate a heart-breaking rugby event. The ball goes up and the crowd holds their breath: will the ball have the legs to make it over the bar? It does, but will drop to the ground only 3.1 metres beyond the posts. Gavin made it, but only just. (Thorburn's famous kick didn't make it beyond the in-goal area.)

For anyone with a healthy kick, who can make the ball travel several tens of metres, $2H/R$ is going to be small. In that case, some approximations can be made. Your maximum range is roughly $L = R - H$. If you can kick the ball 50 metres, the precise formula dictates a maximum goal-scoring distance of 46.9 metres. The approximate formula gives 50 metres – 3 metres = 47 metres, which is fairly close – accurate to within 0.2% percent.

A point to remember is that R goes as the square of the kicking speed. Physics instructs you to put as much oomph into your kick as possible. Place kickers need to develop strong thigh muscles and perfect their technique so that the ball leaves the boot at the fastest speed possible.

And last… If your team is penalised, then not only does the ball need to go over the bar, it has to go over two large gentlemen of the second row, both with arms raised. Ireland could recruit many-capped Matthew O'Kelly at 2.03 metres and Ryan Caldwell, 2.01 to do this. If France's François Tinh-Duc strikes the ball at 45 degrees and can kick it a maximum of 40 metres, then when x=10, y = 7.5. Put another way, even if Messrs O'Kelly and Caldwell stand the regulation 10 metres back with arms raised in the air, the ball will clear their outstretched fingertips by some 4.5 metres. The advantage, therefore, is not in deflecting the ball but obscuring the goal. If the top of the goal post is 10 metres above the ground, then the two-metre tall second rows can block the kicker's view of the posts if the penalty is less than 23 metres from the try line. Such kicks, though, are usually formalities.

Scoring in the corner

There's nothing quite as dramatic as a last-minute try scored right in the corner. The place kicker, though, has the tough job of getting the ball not only over the crossbar, but also between the posts. This is no easy job. The first thing the kicker will do is to move the ball back a certain distance from the try line. It's obvious that this increases his angle, which betters his chance of scoring. Too far back, though, and he won't be able to kick the ball far enough. So, is there a distance L that optimises the chance of scoring? Perhaps there is. The answer requires an excursion into the realm of probabilities.

It would be nice to have a deadly accurate place kicker, such as James Hook. That way, the two points for a conversion are pretty much guaranteed. Suppose instead that you have a fairly poor kicker. The try was scored a distance *d* horizontally from the middle of the posts (see Figure 4.10). Your kicker moves the ball back so that it rests on a little mound a distance *L* from the try line. He aims along the angle *A*, from the mound to the centre of the goalposts. The goal posts are a horizontal distance $d + w/2$ and $d - w/2$ from the kicker. The kicker aims to smack the ball at the angle *A*, where $\tan A = d/L$. If he messes up, and kicks the ball at a different angle, say *B*, he'll still score, as long as $\tan B$ is between

$$\frac{d}{L}\left(1-\frac{w}{2d}\right) < \tan B < \frac{d}{L}\left(1+\frac{w}{2d}\right)$$

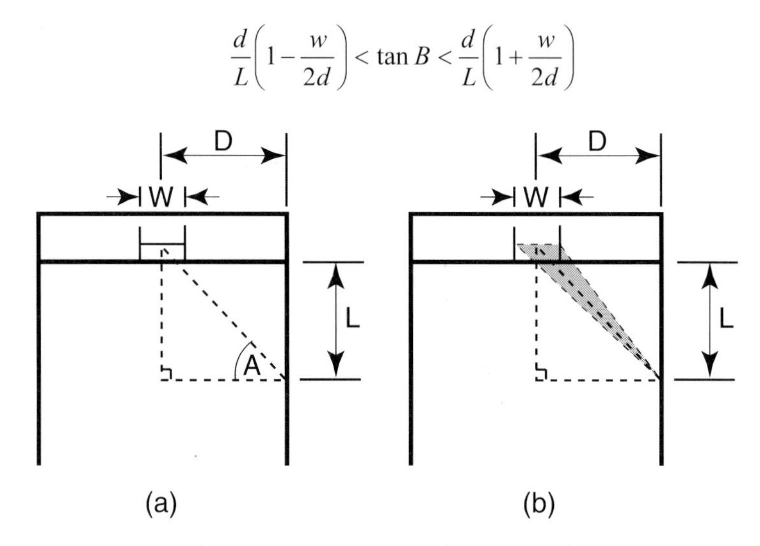

Figure 4.10: A conversion attempt after scoring in the corner.
(a) The ideal kicking angle, (b) The range of successful kicking angles.

The task ahead is to model how good our kicker is. I've played on some teams where the kicker could be relied upon to kick the ball forward, but that was about it. Suppose our kicker is so bad that all we can believe is that the ball will go forward. That's to say, he'll kick it somewhere between straight ahead ($B = 0$) and sideways ($B = \pi/2$). Any angle between these two is equally likely. In that case, the chance $p(B)$ that the ball goes at an angle *B* is *k*, where *k* is a constant. We also know, with absolute certainty, that the ball goes somewhere between 0 and $\pi/2$. Via calculus[10], this means that $k = 2/\pi$.

The chance of scoring, for a poor kicker, is then the fraction of his kicks that head between the angles B_L and B_R, whose tangents are $\frac{d}{L}\left(1-\frac{w}{2d}\right)$ and $\frac{d}{L}\left(1+\frac{w}{2d}\right)$ respectively. The probability of scoring is:

$$P(score) = \frac{2}{\pi}(B_L - B_R)$$

The chance of kicking the conversion is a horribly complicated function of *d* and *L*. One thing, though, is obvious. When you score far out by the touchline, *d* is large and so $w/2d$ is small. When that happens, there's not a lot of difference between B_L and B_R. That means your kicker — who we've modelled as being inaccurate — has only a small range of angles in which to kick and still convert the try: things are not looking good for your team. The kicker, though, does get to decide where to put the ball. That means he determines where *L* ought to be. Calculus, the branch of mathematics that deals with changes, can be of help. It lets us look at how the scoring probability changes when you shift *L*. It shows that the conversion is most likely to be successful[11] if you place the ball at

$$L = d\left[1-\left(\frac{w}{2d}\right)^2\right]^{1/2}$$

This time, the laws of the game, rather than the laws of physics, tell us what to do. The goalposts (Law 1.4a) are 5.4 m apart, which we have called *w*. Likewise, the width of the pitch is also specified, at a maximum of 69 m, so $d = 34.5$ m. From this, the kicker should place the ball about 34 m back. (See Figure 4.11). That means about half way between the 22 and the 10 metre line. Again physics and rugby coincide, for this is roughly where you see the great kickers place the ball. Mind you, our model assumes quietly that there is no connection between range and accuracy. In reality, the larger *L* is the harder the kicker will have to thump the ball and the less accurate he'll become. In other words, there's some dependence on *p(B)* on the distance *L*, a relationship we've ignored. Our results, though, are believable in spite of this flaw.

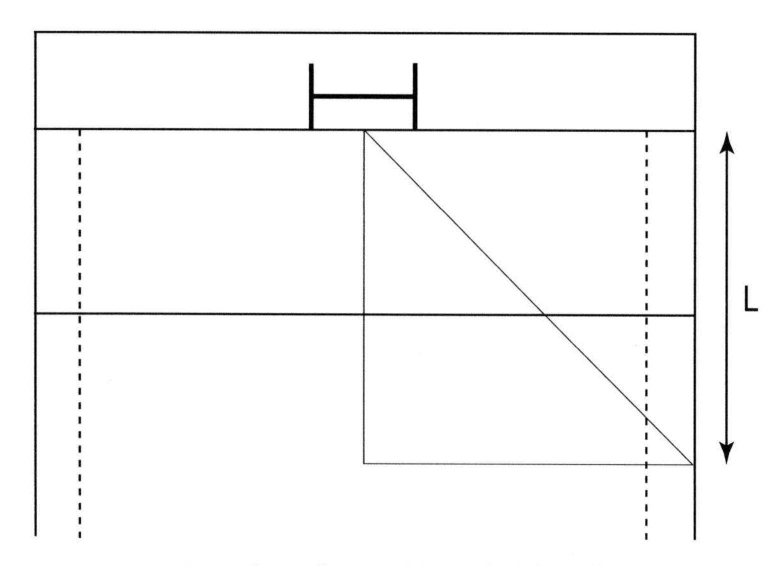

Figure 4.11: The optimal distance back (to scale).

There is one mathematical sting in the tail. Our model assumed we had a really poor kicker, one where any angle between 0 and 90 degrees was equally likely. The physics remains the same even if we have a much better kicker, one who can hit the ultimate angle with an accuracy within, say, plus or minus five degrees. The difference, though, is not in the optimal distance, but in the probability of scoring. If a kicker will hit all angles within the range of 0 and 90 degrees with equal probability, his chance of converting after a try scored in the corner is at most 5%. Our equations show that a great kicker, one who is always within, say, 5 degrees of the intended target, will be on target some 28% of the time. Whether the ball has the legs to get over the bar, though, is another matter.

CHAPTER FOUR - ENDNOTES

1 The ball's momentum, parallel to the ground, remains unchanged. This is, the component of the velocity parallel to the ground is a constant. As the ball is kicked at an angle A with velocity v, the velocity component in the downfield direction is v cos A. So, any time t after the ball has been kicked, it has travelled a distance x down the field, where

$$x = vt\cos A$$

Vertically, things are more complicated. There is an ever-present gravitational acceleration, g that acts straight downwards. The ball is launched with a vertical component of velocity v sin A. It is then immediately grabbed by gravity, which will eventually pulls it back down to Mother Earth. Motion under a constant force is one of the simplest problems in classical physics. The ball is at height y after time t when

$$y = (v\sin A)t - \frac{gt^2}{2}$$

The ball hits the ground when time T has passed, with

$$v\sin A = \frac{gT}{2}$$

At this time, the ball will be a horizontal distance R, its range, from Traille's boot. To find this distance R, we put the equation for T into the equation for x. It gives:

$$R = (v\cos A)T = (v\cos A)\frac{2v\sin A}{g} = \frac{v^2\sin 2A}{g}$$

because, for any angle A, $\sin 2A = 2\sin A\cos A$.

In terms of height, the ball will be its maximum distance above the ground when it has gone halfway through its journey. This occurs at a time $t = v\sin A/g$ Put this in the expression for y to get:

$$H = v\sin A(\frac{v\sin A}{g}) - \frac{1}{2}g(\frac{v\sin A}{g})^2 = \frac{v^2\sin^2 A}{2g}$$

2 The ball bounces again (i.e., its height is zero) after a time $2T$ given by $2T = 2ev\sin A/g$. In that time, the ball has gone a distance $X = 2vT\cos A$. So, during the time the ball

has gone from the moment of its first bounce to the moment of its second bounce, it has travelled a distance:

$$X = \frac{ev^2 \sin 2A}{g}$$

So, at the end of the first bounce, the ball has travelled a total distance *R(1),* where

$$R(1) = \frac{v^2 \sin 2A}{g} + e\frac{v^2 \sin 2A}{g}$$

This can be written more cleanly as:

$$R(1) = (1+e)\frac{v^2 \sin 2A}{g}$$

After the second bounce, the ball's horizontal speed is still $v\cos A$. Its vertical speed is changed, though. The muddy surface has sucked more energy out of the ball, so that now it heads upwards once more but with a new speed, $e^2 v\sin A$. This is exactly the same situation we were in after the first bounce, only with *e* replaced by e^2. This lets us use the same results; all we have to do is replace *e* by e^2 in our final equations. What's more, we can repeat this over and over again. So, after *N* bounces, the ball will be a distance *R(N)* from where it was kicked, with $R(N) = (1+e+e^2+\cdots e^N)\frac{v^2 \sin 2A}{g}$.

3 The equation of an ellipse is $\frac{x^2}{a^2}+\frac{y^2}{b^2}=1$, or equivalently $y^2 = b^2\left(1-\frac{x^2}{a^2}\right)$. A small element of the ellipsoids volume is $dV = \pi y^2 dx$, which is $dV = \pi b^2\left(1-\frac{x^2}{a^2}\right)dx$. The total volume of the sphere is the integral of this from –*a* to +*a*, or twice the integral from 0 to a. Hence:

$$V = 2\pi b^2 \int_0^a \left(1-\frac{x^2}{a^2}\right)dx = 2\pi b^2\left[x-\frac{x^3}{3a}\right]_0^a = \frac{4\pi}{3}ab^2$$

4 The actual expression for the surface area is

$$A = 2\pi\left[b^2 + b\sqrt{a^2-b^2}\,K(k) + \frac{b^3}{a^2-b^2}F(k)\right]$$

Here $k = \arccos(b/a)$ while *K* and *F* are the complete elliptic integrals of the first and second kind.

5 Conservation of angular momentum requires:

$$mLv + \frac{ML^2}{3}\Omega = -mLu + \frac{ML^2}{3}\omega$$

After kicking, the ball exits stage left with a new speed *v* while Johnny's leg now rotates about the hip with a new frequency Ω. The equation reduces to:

$$\Omega = \omega - \frac{3m}{ML}(v+u)$$

Kinetic energy is also conserved. The total kinetic energy before the collision is:

$$\frac{1}{2}mu^2 + \frac{1}{2}\left(\frac{ML^2}{3}\right)\omega^2$$

while the energy after the collision is:

$$\frac{1}{2}mv^2 + \frac{1}{2}\left(\frac{ML^2}{3}\right)\Omega^2$$

Only a fraction *e* is transferred into the ball's kinetic energy after the kick.

The equations for kinetic energy need to be applied in a so-called reference frame that moves with a speed *u*. There is a lot of tedious algebra to go through before ending up with the equation in the main text.

6 The sine function goes up and down as the angle changes. Sin 0 is 0 and sin 90 = 1. Beyond 90 degrees, sin decreases in value, but remains positive. By the time the angle reaches 180 degrees, the sine function has returned to zero. There is, therefore, a value of 2A, bigger than 90 degrees, such that sin 2A = ½. To find it, set 2A = 90 + 2B. We seek $\sin(90+2B)=0.5$. However, $\sin(90+2B)=\sin 90\cos B+\cos 90\sin B$ and, as sin 90 = 1 while cos 90 = 0, we require cos 2B = ½. The solution to this equation is 2B = 60. As 2A = 90 + 2B, this means 2A = 150, or A = 75 degrees.

7 As a plausibility argument, we can obtain this by usage of Bernouilli's principle, which says that the pressure in a fluid that moves with speed *u* is $p+\frac{1}{2}\rho u^2 = const.$ Suppose that the wind blows with speed v and the ball rotates with angular velocity ω. Roughly speaking, the pressure on, say, the top half of the ball is $p_{TOP}+\frac{1}{2}\rho\,(v-R\omega)^2 = const$.

The pressure on the bottom half of the ball is $p_{BOTTOM} + \frac{1}{2}\rho\,(v + R\omega)^2 = const$ and so the pressure difference between the two is approximately $\rho v \omega R$. Force is pressure multiplied by area. As the half the surface area of the cylinder is πRL, the Magnus force amounts to $\pi \rho v \omega R^2 L$. The direction of the force depends on which way the wind blows and which way the ball rotates.

The simplest mathematical way to obtain the Magnus force, or the Robins-Magnus force, is to resort to the motion of a spinning cylinder in an incompressible, irrotational flow. The lift force per unit length is $\rho v \Gamma$, where Γ is the circulation. The circulation for a cylinder of radius R spinning with angular velocity ω is $\Gamma = \pi \omega R^2$. Combining, the Robins-Magnus force on a cylinder of length L is $\pi \rho v \omega R^2 L$.

There are more formal ways to obtain the same result. One way is to look at the lift force on a rotating cylinder in an incompressible, irrotational flow. Another is via the Kutta-Joukowski theorem for the lift on an airfoil.

8 The range is:

$$x = \frac{v^2}{g}\sin 2A\left(1 - \frac{u}{v\cos A}\right) = \frac{v^2}{g}\left(\sin 2A - 2\frac{u}{v}\sin A\right)$$

which uses the double-angle formula for the second term. This can now be differentiated with respect to the angle A. The range is maximum when this derivative is zero. That is to say, when:

$$\frac{\partial x}{\partial A} = \frac{v^2}{g}\left(2\cos 2A + 2\frac{u}{v}\cos A\right) = 0$$

We now write $\cos 2A = 2\cos^2 A - 1$, to give us a quadratic expression:

$$2\cos^2 A + \frac{u}{v}\cos A - 1 = 0$$

The solution to this is $\cos A = \frac{1}{\sqrt{2}}\left(1 + \frac{u^2}{8v^2}\right) + \frac{u}{4v}$

9 We can adapt some physics described by Carl Mungan in his article for *The Physics Teacher* Volume 41, 132 (2003). Life is simpler to say that the ball is launched from a height -*H* at speed *v* at an angle *A*. Then the height is:

$$y = -H + (v\sin A)t - \frac{1}{2}gt^2$$

above the ground after time t. At the same time, it is a distance x down the field, where

$$x = (v\cos A)t$$

We know that the ball is at height 0 when it has travelled a distance L. So, we can combine the equations to obtain:

$$-H(1+\cos 2A) = \frac{gL}{v^2} - L\sin 2A$$

To find the optimal launch angle, set $dL/dA = 0$, which gives the condition $\tan 2A = -L/H$. From this, we know that

$$\cos 2A = \frac{H}{\sqrt{H^2 + L^2}}$$

while

$$\sin 2A = \frac{L}{\sqrt{H^2 + L^2}}$$

This is what we need. Our equation involving cos 2A and sin 2A now reduces to:

$$H = \frac{v^2}{2g} - \frac{gL^2}{2v^2}$$

If the maximum range you can kick the ball is R, then $R = v^2/g$. You score if provided $L^2 = R^2 - 2RH$, or $L = R\sqrt{1-(2H/R)}$. For small x, $\sqrt{1-2x} \approx 1 - x + \cdots$, so that $L \approx R - H$.

10 Suppose the probability $p(B)$ of the ball heading at an angle B is k and this doesn't depend on angle. The chance that the ball heads in a direction between angle A and C is:

$$prob(A \le B \le C) = \int_A^C k dB$$

If we know with certainty that the ball heads between 0 and 90 degrees, then:

$$1 = k\int_0^{\pi/2} dB = k\frac{\pi}{2}$$

This means that $k = 2/\pi$.

11 The probability of scoring is $p = \frac{2}{\pi}\left[\arctan\left(\frac{d}{L}\left\{1-\frac{w}{2d}\right\}\right) - \arctan\left(\frac{d}{L}\left\{1+\frac{w}{2d}\right\}\right)\right]$. To find the value of L that maximises p, form:

$$\frac{dp}{dL} = 0 = \frac{d}{dL}\left[\arctan\left(\frac{A}{L}\right) - \arctan\left(\frac{B}{L}\right)\right]$$

where A and B are shorthand for the complicated terms in the previous equation. To make life simpler, we can set $x = 1/L$. The probability will then be a maximum if:

$$x^2 \frac{dp}{dx} = 0 = \frac{d}{dx}\left[\arctan(Ax) - \arctan(Bx)\right]$$

The derivative of $\arctan Ax$ with respect to x is something we can look up in mathematics tables. The answer is $A/(1 + Ax^2)$. The condition for the probability to be maximised is then:

$$\frac{A}{1 + Ax^2} - \frac{B}{1 + Bx^2} = 0$$

the solution of which is $x = 1/\sqrt{AB}$. As $x = 1/L$ and $A = d\left(1 - \frac{w}{2d}\right)$ while $B = d\left(1 + \frac{w}{2d}\right)$, this predicts that the optimal place to put the tee on the turf is at

$$L = d\sqrt{(1 - \frac{w}{2d})(1 + \frac{w}{2d})} = d\sqrt{1 - \left(\frac{w}{2d}\right)^2}$$

Chapter Five

Match day: On the weather you'll play in and whether you'll win

'Do you know what? I'd give it all up tomorrow, the whole lot, for one Irish cap. Just one. There is hardly a day that passes that I don't think of what it would be like to run out on to Lansdowne Road as one of the Irish team.'

Richard Harris

'Look what these #&*@ have done to Wales. They've taken our coal, our water, our steel. They buy our houses and they only live in them for a fortnight every 12 months. What have they given us? Absolutely nothing. We've been exploited, raped, controlled and punished by the English - and that's who you are playing this afternoon.'

Phil Bennett's pep talk

There comes a time when practice ends, the pep talk is over, and real rugby begins. There is a team to defeat, a league or cup to be won, a wooden spoon to be avoided. Skill and teamwork help you to victory; blind chance, though, has a role to play. So, too, does Mother Nature. In the end, the number of points you will put on the score board by the end of the game is anyone's guess, as bookmakers know all too well.

Growing hot and cold

There's a world of difference in playing sevens in Dubai (about 25 Centigrade) and playing the full fifteen-a-side game in winter at Croke Park. Some people cope better in the heat. Others thrive when the weather turns cold. Biologists and physicists have teamed up to explain why. The key is how much energy a player takes in, and how much he loses.

The energy anyone takes in – by eating and drinking – is related to their mass. Bigger animals eat more than smaller animals. If nothing else, the stomach of an elephant is larger than a squirrel's, so the elephant takes in more food. Put another way, after the game, Ireland's Tony Buckley, the heaviest man in Six Nations Rugby history at 134 kg, is likely to tuck away more food than his 80kg compatriot, scrum half Eoin Reddan.

Isaac Newton, he of laws of motion fame, explored how objects cool – how they send heat energy into the world about them. The rate at which an object gives off heat, he said, depends on the excess temperature. The higher your temperature relative to the outside air temperature – the excess temperature -- the more energy you pump into the atmosphere. A thermometer in your mouth, under your armpit, or somewhere unmentionable, displays a temperature of about 98.4 degrees Fahrenheit, or 37.4 Centigrade. Americans run a little higher, at around the 98.7 degrees mark. It's a hot day indeed where outdoor temperatures are higher than this. (It can happen: back in 2007 when the London Welsh played Portugal in Vilamoura as a 'warm up' for the world cup, it was 38 degrees; the away team were victorious, 16-19).

Normally, the inside of our bodies are usually warmer than the outside of our skin. It's a fact of life – enshrined in the second law of thermodynamics -- that heat flows from a hot object to a cold one. So, heat usually flows from your body out into the atmosphere.

A French scientist, Joseph Fourier, also studied heat. Fourier, an advisor to Napoleon, was keen to cool off the Emperor's cannon. Fourier's law suggests that the energy a hot body pumps into the cooler air around it depends not only on the excess temperature but also on the surface area of the object. For biologists, this means that an animal loses energy depending on its surface area.

What's true for objects and animals is true for rugby players. Suppose every member of the team is a solid block of stuff. Each player has a height H and, from above, looks like a square whose side is of length L. The bigger the block, the more the player weighs, the bigger his stomach, and the more fuel he needs to survive. In other words, the energy he takes in depends on his volume, which is HL^2. Our solid block of rugby prowess loses energy across his surface area. But the surface area of a cube consists of the area of the sides (four lots of HL) and the top and bottom (two lots of L^2).

The ratio K of the heat energy lost to the energy consumed is the ratio of area to volume:

$$K = \frac{area}{volume} = \frac{4HL + 2L^2}{HL^2} = \frac{4}{L} + \frac{2}{H}$$

Now suppose the weight of the player is fixed. That means, roughly speaking, the volume V is a constant. As $V = HL^2$, which is constant, $L = \sqrt{V/H}$. The surface area to volume ratio is then:

$$K = 4\sqrt{\left(\frac{H}{V}\right)} + \frac{2}{H}$$

Yet again, calculus lends a hand[1]. We can work out what value of H gives the minimum value of K. The answer is $V = H^3$. As $V = HL^2$, the minimum K value is for players close to $H=L$, i.e., if two players have the same weight, the one with the lowest K value is the one who most looks like a cube[2].

Noncubic players will have a higher value of K than their cubic counterparts. A prop may be roughly cubic, but a second row forward may be half a metre taller and yet have the same weight. In that case, as the two players' weight and volume are the same but the lock will have a bigger value for K.

This is the key to playing at extreme temperatures. Suppose it's frigid. You want to give up as little heat to the atmosphere as you possibly can, so you want a low surface-to-volume ratio, i.e., a low value of K. The players who are mostly cubic will do best.

The opposite is true in hot climates, when playing in the Dubai sevens. There, to avoid overheating, you want the largest K you can get. A large K value is for the skinny creatures. The backs should be fine, but props are going to suffer terribly in the heat. In a sevens tournament, played in summer under a hot sun, you might want your three-man scrum to consist of the hooker and those who play flanker in the fifteen-a-side game. They shouldn't keel over with heat exhaustion before the first half has finished.

If we truly think of players as cubic blocks, the world wide web can help us out by providing some intriguing numbers. Japan's try scoring phenomenon Daisuke Ohata is 1.76 m tall and tips the scale at 80kg. By contrast, Welsh tight-head prop Adam Rhys Jones is 1.83m tall and has a mass of 127kg.

To work out their K values, we need to convert their mass into a volume. As density is the ratio of mass to volume, all we need to know is the average density of a human. Luckily, humans can just about float in a pool or in the ocean. That means we must have about the same density as pure water, which is 1,000 kg/m^3 , and salt water, which is 1,025 kg/m^3. Either value would be good enough for our purposes, but oddly enough the typical human density is almost exactly between these two values, at about 1,010 kg/m^3.

Armed with this, we write:

$$K = 4\sqrt{\left(\frac{H}{V}\right)} + \frac{2}{H} = 4\sqrt{\frac{1010 \times H}{M}} + \frac{2}{H}$$

where to get the numbers right we now have to enter height in metres while mass, M, is in kilograms. Putting in the numbers:

$$K(Daisuke) = 4\sqrt{\frac{1010 \times H}{M}} + \frac{2}{H} = 4\sqrt{\frac{1010 \times 1.76}{80}} + \frac{2}{1.76} = 20$$

while:

$$K(Adam) = 4\sqrt{\frac{1010 \times H}{M}} + \frac{2}{H} = 4\sqrt{\frac{1010 \times 1.83}{127}} + \frac{2}{1.83} = 16.4$$

The scrum half puts out about 18% more heat energy for his weight than Adam does. Daisuke will feel the cold, Adam will not. Jones will suffer on hot days, such as the EDP Final in 2007 at Twickenham, where his Ospreys team fell victim to Harry Ellis and the Leicester Tigers. A year later, though, when the temperature was slightly cooler, Jones's revenge was complete[3].

WHEN IT'S COLD OUTSIDE

Every other year, the Welsh play in Edinburgh in the Six Nations tournament. It's a cold day in Auld Reekie and a savage wind blows. Dwayne, Shane, and the rest of the Welsh backs will suffer because of their high surface-to-volume ratio. Their bodies will try to help them out, and the ways they respond obey sound physical principles. We are warm-bloodied creatures.

When warm blood flows through capillaries and other blood vessels close to the surface of Welsh skin, it releases heat energy out into the cold Scottish atmosphere. This makes Edinburgh warmer and the Welsh players colder. It's also the operating principle behind a car radiator, where water circulates around the engine block; the water heats up, cooling the engine off, but in the radiator the heated water gives up its energy to the outside world, cooling itself down in the process. Blood flowing near the skin surface is like water trickling through the radiator fins. To staunch his loss of heat energy, Shane's body constricts these blood vessels, shutting off blood supply to his extremities. A Caucasian is usually some variety of pink and pink, this, as we all know, is a combination of red and white. If blood isn't flowing close to the surface of your skin, then you are short of red. So, people who are cold look less pink or more white. If Dwayne Peel gets even colder, then his veins, which carry deoxygenated blood, become more prominent. Venous blood is a much darker red, almost purple. When seen through a veil of white flesh, he will look blue.

When the blood supply to the fingers and toes shut down long enough, you risk frostbite. As a first step, fingers become numb. That's why halves and three quarters often drop the ball on cold days, or constantly commit knock ons. It's not incompetence, it's merely the early stages of frostbite, the warning signs of which are the umbles — stumbles, fumbles, mumbles, and grumbles. Your body tries to keep warm by shutting down the flow of warm blood, leaving your limbs (fumbles and stumbles), lips (mumbles) and brain (grumbles) a bit short of O_2.

Frostbite might not seem like a serious possibility — unless you're one of those hardy souls who've played rugby in Antarctica! Each South Pole summer sees a team of American scientists from McMurdo Station (the Mount Terror RFC no less) take on one from Scott Base, home to a crew of New Zealanders. Imagine having to stand still in Antarctica and freeze while watching the haka. This pre-game chill seems to work – the Americans have never won. And as Antarctica hosts a 15-a-side game, it's natural that the Arctic should focus on Sevens. The Finnish ski resort of Saariselkä, up in Lapland way beyond the Arctic Circle, is home to the annual Arctic Rugby Tournament where, in 2008, the Helsinki Warriors conquered all.

When you exercise, your body needs more oxygen and so your heart pumps faster and harder. The increased pumping forces blood into the outermost capillaries. This counteracts your body's desire to clamp down those blood vessels and makes it pump warm blood wherever you may need it. In cold weather, you might not feel like moving, so your body forces you to move: you shiver. Before you get to such a stage, run around! If you're a back and you're not getting much of the ball, stamp your feet. Yet again, the laws of physics help to explain why we do what we've always done.

In warm weather, your body does the opposite. Adam Rhys Jones needs to lose more heat than usual, so his heart pumps more blood close to the skin's surface. He's usually a pale shade of pink, but when more red blood flows near to his skin surface, he'll turn a variety of beetroot. The heat is transferred from your blood vessels, through your skin, into the air.

Your body does something else to keep you cool. If you supply enough heat energy to a solid, it melts. Supply enough heat energy to a liquid and it becomes a vapour. One of the differences between a liquid and a vapour is the average distance between the molecules. Molecules in a gas are, on average, farther apart, which is why gases are less dense than liquids. When you supply heat energy to a liquid, some of it heats the liquid up. Another part transforms the liquid into vapour. There's a certain, definite amount of energy, L, needed to convert 1 gram of liquid into 1 gram of vapour. This is the latent heat of vaporisation. For water, it is about 540 calories, or roughly 2,300 Joules, per gram. Rugby players use the benefits of latent heat a lot: they sweat.

Sweat turns out to be one of our most useful strategies. Humans can't run fast, not compared with animals. We can, however, outlast them. That's possibly how our Cro-Magnon forebears hunted – just keep running and tire the animal out. We are relatively hairless and can sweat, which means we don't have problems with overheating; animals do. An antelope, for example, is covered with a layer of hair so sweating is not really an option for a springbok (the animal) compared to a Springbok (also an animal, of course, but slightly more intelligent). The animal has to pant to cool down, and it's tough to run and pant at the same time. To show how much we are built for endurance, the oldest 100 mile race for humans, the Western States Endurance Run, has a record time of 15 hours, 36 minutes, and 7 seconds, which was set in 2004 by Scott Jurek. The Tevis Cup, a 100 mile race for horse, has a best time of 10 hours 46 minutes. When you bear in mind that in 2008 Midnight Lute won the Breeder's Cup Mile in a time of 1 minute 7.08 seconds while the human record for the mile stands at 3 minute 43.13, you can see how much closer we are to our equine friends over the longer distance.

Suppose, in one second, Adam Rhys Jones gives off a certain amount of heat energy, call it *E*. It takes 2,300 Joules to turn one gram of water into vapour, so *E* Joules of energy will transform *E*/2300 grams of sweat into water vapour. If Adam, an elite athlete, generates a power of about 500 Watts, then in the course of an 80 minute game, he'll generate some 500x60x80 = 2.4 million Joules of energy. As it takes 2,300 Joules to get rid of one gram of sweat, Adam can evaporate about 1,000 grams of sweat during the game. This is 1 kilogram of salty wetness or, for the virulently anti-metric, about 2 pounds, which is why your weight can drop significantly by the end of a strenuous match. Physics suggests two things. First, replace that water, so you can avoid dehydration by continuing to sweat (after all, no-one wants a prop to have heat stroke). Second, sweat is not water alone, but water, salt, and trace elements combined. At half-time, replace the water you've lost, but the salt and electrolytes, too. This is particularly true for anyone with cystic fibrosis which, among other things, makes your body pump out far more salt in your sweat. Fortunately, there are many drinks on the market to help you replace all the fluids and goodness you've lost during the first 40 minutes of the game. If not, the traditional half-time slices of orange will, as veteran BBC commentator Bill McLaren might say, do the needful.

The water that evaporates off your skin cools you down. That's why, even on a warm day, you can feel cool when you get out of the shower. The water droplets suck energy from your body and then evaporate into the air in your bathroom. Air is a subtle blend of nitrogen and oxygen, for the most part, but also water vapour. Water not only evaporates off of your body, but also from the surface of lakes, rivers, and streams. So, air is seldom truly dry. There's only a certain amount of water, though, that the air can hold at any given temperature. If the water in the air is, say, 70% of this maximum value, then the relative humidity is 70%. The more water is in the air, the tougher it is for sweat to evaporate. On muggy days -- when the relative humidity is pretty high -- your sweat will soak into your jersey. The cooling mechanism won't work; you just end up feeling hot *and* sticky. You'll definitely need a jersey that can absorb moisture well. Bring two - one for after half-time.

Your body has mechanisms to control your temperature, but there are other things you can do to heat up or cool down. Again, physics leads the way. Heat loss, or heat gain for that matter, comes in three flavors: conduction, convection, and radiation. There are strategies to cope with all three.

Suppose you apply heat to the bottom of an iron rod. Given enough time, the top of the rod gets hot, too. No metal moves; it's just that one layer of metal heats the layer of metal above it, which in turn heats the layer of metal that lies next to it, and so forth. This process is called conduction and physics can deal with it fairly easily. Suppose a rugby player is standing still – not that he should. He's been running around and getting hot and this heat energy leaves his body at a rate of W watts and passes into the thin layer of air between him and his jersey, heating the trapped air to the same temperature as his skin. The temperature on the inside of the jersey is T, but on the outside it'll be the same temperature as the cold midwinter afternoon, T_{OUT}. If the jersey is made of a material of thickness d, then the law of heat conduction says that:

$$T = T_{OUT} + \frac{Wd}{Ak}$$

Here k is the thermal conductivity of the jersey, which is a constant that depends on the material and the manufacturing process. (See Figure 5.1)

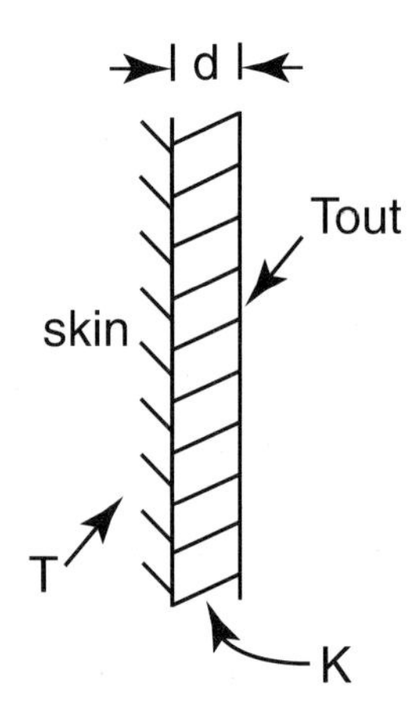

Figure 5.1: The rugby jersey, thickness d and conductivity K.

The temperature at Murrayfield, T_{OUT}, is not something Scotland international Mike 'Blade' Blair can control. To keep warm, Mike needs T, the temperature on the inside of his shirt, to be as high as possible. Physics requires him to put on a thick jersey (to increase d), one that fits reasonably well, made of a good insulator (low k), and covers a large area of his body.

For a long period of history, physicists thought that heat was some form of liquid called the caloric. We now know that's not the case, but the analogy can prove useful. Suppose you have a liquid that's being produced at a rate S. The concentration of this liquid, C, will obey the same equation as the one for heat conduction. In other words, if C is the concentration of sweat on the inside of your jersey, then:

$$C = C_{OUT} + \frac{Sd}{AD}$$

where D is the diffusivity of the liquid (sweat) in the substance (the jersey). Unlike the temperature equation, we want the concentration of sweat next to your skin, C, to be as small as possible. This means you want a high value for D.

Manufacturers have developed special thermal manikins that can be heated up when clothed, so that the material's thermal conductivity can be measured. The thermal conductivity of cotton is about 0.03 W/m K, well short of the value for a metal. Research shows, though, that the thermal conductivity of cloth increases when it's wet. That's to say, if C increases, T decreases. If he sweats profusely, Blair's skin temperature is going to go down. To avoid this, a dry shirt for half time would come in useful, or a jersey made of a special wicking fabric. (One such jersey has recently been selected by the Welsh national team; it's made by Under Armour, a company based here where I write, in Baltimore, USA, home of the Baltimore-Chesapeake Brumbies. Scientists at Cornell University have produced a 29-page document singing the praises of the Under Armour cloth and its wicking properties. Its diffusivity is about 100 times greater than that of regular cotton.)

As far as your own rugby kit goes, there are a couple of items on the agenda. First, dress in layers to keep warm. This is the equivalent of double glazing, for it traps a layer of air (a notoriously poor conductor of heat) between the layers of cotton. You could wear two pairs of socks to double the thickness of the insulation. Buy a thick jersey, not a cheap imitation one and/or wear a T shirt underneath your jersey.

The conduction equation works best when every inch of you is covered by a protective layer of cloth. In reality, you keep warm by covering your body's surface area by as much cloth – insulation – as possible. So, keep your socks rolled up, your collar up (if your jersey has one), and pull your sleeves down to cover your wrists. Last, the equation assumes that the cloth material is perfectly smooth. It isn't. Holes in the material will let heat energy pass through and ruin the insulating effect. To reduce the number of holes in the cloth, a densely woven material is best. And, of course, we can insulate ourselves. A lavish covering of body hair will help provide a thermally insulating barrier to the outside world and help trap warm air close to your skin. As exhibit A, see Sale Sharks and France's Sébastien Chabal.

For forwards only, an old-fashioned scrum cap, or some bandage wrapped around your head and held fast by duct tape, will also keep you warm – ears have a large of surface area through which you can lose heat energy and duct tape is a pretty good insulator. A scrum cap also makes your ears less appetizing to the opposition's front row.

On a warm day, or when you're playing sevens, the opposite holds true. Clothe yourself in a loose-knit rugby jersey. This should be short sleeved, or at least you should roll the sleeves up. Short socks, or socks rolled down, will help you to keep cool as well.

HUBBLE, BUBBLE, TOIL AND TROUBLE

Hot air rises and cold air sinks. That's the basic premise underpinning convection. As you heat up the air next to your skin it becomes warm. Warm air is less dense than cold air, and – like a hot-air balloon, this air wants to leave your skin and float away. Disturb the warm air slightly and it floats off, taking some heat energy with it. It is replaced by cold air. The net result: you feel cooler.

On frigid days, your body fights back. Your hairs will stand on end and you'll get goose bumps. There's a reason for this. Goosebumps make the surface of your skin much rougher than normal. This makes it more difficult for that warm layer of air to break off from your skin and head on into the environment. The hairs serve as a barrier to prevent the so-called thermal boundary layer from taking flight. So, in addition to its conduction method, the body also has a method for reducing the amount of heat you lose via convection. If you get too hot, then, roll up those sleeves and allow that hot air, which was trapped between your body and your jersey, to take heat energy away. To be really daring, you could shave your arms and legs, as this will allow sweat to form and evaporate more easily, leaving you cooler but balder.

GETTING FRIED

The Sun is the closest star to Planet Earth. Our stellar neighbor is mostly a mixture of hydrogen and helium, elements that were forged early on in the universe in the Big Bang crucible. At the centre of the Sun, the pressure and temperature are phenomenally high. So high, in fact, that a process unknown on Earth can take place. In a hydrogen atom, a negatively charged electron orbits the positively charged, and much more massive, proton. Solar temperature and pressure are so high that the electron is stripped away from the proton. Then protons are rammed together, two at a time, along with two (electrically neutral) neutrons, to form the nucleus of a helium atom. One curious thing about the helium nucleus is that it is less massive than the four particles used to construct it. This so-called mass deficit is why the sun shines.

In 1905, the soon-to-be-famous Albert Einstein launched what later became known as special theory of relativity. In what amounted to an addendum, he looked at what happens to energy when you travel close to the speed of light. In that article Einstein proposed, in words, a formula that has since become famous throughout the world. He wrote it down for the first time as an equation in 1907, as $L = mc^2$. The L was then crossed out and replaced with by E. In the Sun, because the helium atom has a mass deficit compared to the two protons and neutrons that went to make it, it gives off an energy E equal to the mass deficit multiplied by the speed of light. $E = mc^2$. The sun pumps out energy about 4×10^{26} Joules every second and, as the speed of light is 3×10^8 m/s, this suggests the sun is on a crash diet where it loses about 5×10^9 kilograms every second. Fortunately for us, the sun is no lightweight. Although it would float is a cosmic bowl of water, the solar sphere has a mass of 2×10^{30} kg, so it should be around to shine on rugby teams for the next ($2 \times 10^{30}/5 \times 10^9$) seconds, which is about 10^{10} years.

Solar energy whizzes towards Twickenham through about 8 light-minutes of interplanetary space. There is neither any solid to conduct this energy, nor any liquid or gas to convect it. Instead, the Sun's energy gets to us via radiation. This is perhaps not as easy as it sounds. Einstein, in another of his famous papers from the year 1905, showed that light could sometimes behave like waves and sometimes like particles. When you try to balance the books of the fusion process, with the protons becoming helium nuclei, it's best to think of the energy as a collection of high-energy particles of light. As the photons travel towards the surface of the sun, they scatter off other particles and molecules, losing energy with each encounter. Closer to the surface of the sun, this energy is best thought of as waves, rather than as particles. Just like waves on the surface of the ocean, there are crests and troughs, amplitudes, frequencies, wavelengths, and so forth. From the Sun, some radiation has wavelengths suited to the human eye. This kind of radiation we call light, and it encompasses all the colours of the rainbow. Some radiation has slightly longer wavelengths, such as the infrared, while others have shorter wavelengths, including ultraviolet. Infrared radiation will heat us up; ultraviolet radiation, although partially blocked by our atmosphere, can cause skin cancer. (Make sure you wear sun screen with a high SPF. This prevents melanoma, makes you more slippery, and thus more difficult to catch). In the end, the Sun bathes the Earth's surface with radiation whose power is about 1400 Watts per square metre.

On a cold day, we are warmer than our surroundings and emit heat energy in the form of radiation. The power, W, with which this radiation is emitted obeys the Stefan-Boltzmann law. As this energy has to pass through your jersey first, the power of the radiation emitted is:

$$W = Ae\sigma(T^4 - T_{OUT}^4)$$

where T is your skin/jersey temperature, T_{OUT} is the temperature of your jersey, and σ (the Greek letter sigma) is one of the most fundamental numbers in all of physics, Stefan's constant. As usual, A is the surface area, e is the emissivity of your jersey and is a measure of how readily it radiates. The value of e for human skin is about 1, which means that humans are close to being perfect absorbers and emitters of radiation. The emissivity of cotton, the cloth from which your rugby jersey is probably made, is about 0.77.

On a cold day, you want to reduce the amount of energy you radiate into the atmosphere. The key part of the equation is the term Ae. Suppose that a fraction of your body area, f, is bare skin of emissivity 1, while the rest of your body (a fraction $1-f$) is covered in cloth of emissivity e. Then $Ae = A\left[f + (1-f)e\right] = A\left[f(1-e) + e\right]$. As e is somewhere between 0 and 1, the way to minimise Ae, and so reduce the energy you radiate, is to make f as small as possible. In other words, the more of your body that is covered in cloth, the warmer you'll be. There is a team decision to make, however. The team's jerseys, shorts, and socks can help or hinder the quest to remain at the perfect temperature. Dark colours have different emissivities from light colours. So, if you wish to keep warm in winter, follow the model of Neath, Saracens, and the All Blacks. If you're playing in summer, make sure your sevens outfit is all white, just like England's or Ulster's.

One additional thought on colours. Anthropologists at Durham University in England have shown that, in matches that pitch a blue team against a red team, the red teams win far more frequently. The researchers claim that red is a psychologically intimidating colour. So, during the Six Nations tournament, the Welsh are warmer, more intimidating, but perhaps less stylish than the Italians. No-one gave the anthropologists' message to Stade Français, who wear pink and blue. Their September 2008 game against Toulon (who sport red and black, the colours of blood, death, and the devil) was marked by a rolling brawl that lasted several minutes, which earned red cards for two props, a 20-day ban for the Toulon scrum half Norman Jordaan, and fines of 5,000 Euros for both teams. Not much intimidation there!

The state of the pitch

After you've use fashionista physics to dress appropriately for the weather conditions, the next thing you may notice is the state of the pitch: wet, dry, flat, or bumpy are just some of the options. The game depends crucially on what the ground is like. A hard-baked pitch will have a high coefficient of restitution. A ball that lands in front of you is likely to bounce high and go far. In wet conditions, the ball will die where it lands and everyone will have trouble with traction. If the topsoil is muddy, get a longer stud if you want to push in the scrum or sprint effectively. (I once played in a game where both the locks had forgotten their boots, if you can believe it, so wore training shoes instead. The pitch was wet to begin with and, as the game wore on, it became increasingly cut up and muddy. It was hell in the front row; it's almost impossible to hook the ball when you're team is being pushed back almost at running speed.)

The state of the pitch, in terms of physics, is governed by the same set of laws as the heating of humans. The sun warms the surface, at a rate of *R* Watts per square metre. This energy converts the water in the topmost layers of grass and soil into water vapour. If these were the only processes around - and they are not - we can balance their effects. If the ground neither heats nor cools during the game, this solar energy is devoted to evaporating a thin layer of water. (See Figure 5.2) The depth of water *h* that evaporates every second when the latent heat of vaporisation is *L* Joules per kilogram is:

$$\rho L h = R$$

in which ρ (the Greek letter rho) is the density of water.

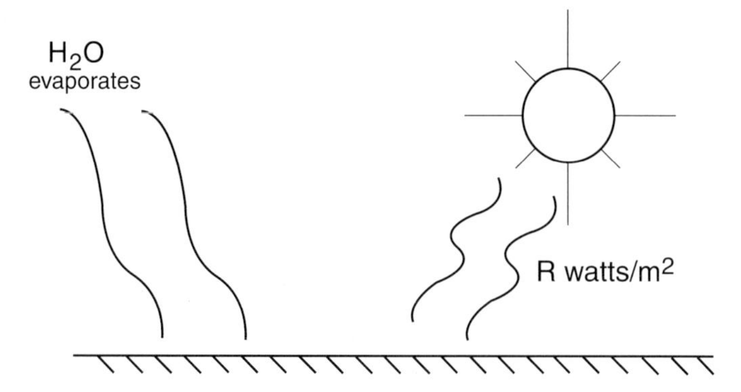

Figure 5.2: The sun's radiation provides enough energy to the water in the soil to convert it to water vapour.

If the turf is a luxuriant green and reflects back most of the incident sunlight, R is about 500 Watts per square metre. The density of water is 1,000 kg per cubic metre, and the latent heat of vaporisation is about 2.5 million Joules per kilogram. So, the sun has the capacity to evaporate a layer of water, of thickness about 20 x 10^{-8} metres, every second. That's tiny, about the same as the wavelength of red light. In a day, though, it adds up. In 24 hours, the ground gives up about 1.75 cm or about 0.7 inches of water. To keep the turf in good shape, a groundskeeper needs to replace at least this much water every day, either through rain from the heavens above or from a sprinkler system. If not, then – unsurprisingly – the soil becomes parched. After a few rainless days, the surface will be devoid of water and this arid region will be deeper than the roots of the grass. The grass will wither and ruin the playing surface.

The rate at which the water is removed from the soil depends on how strong the sun is, whether the days are humid or dry, and whether or not a stiff breeze is blowing. Of course, you might want to have a quiet word with the local grounds staff. If you have some very fast backs and a fullback who can kick the ball far, you might benefit from a drier pitch. On the other hand, a mighty pack might like to grind out the full 80 minutes in scrum, ruck, and maul on a swamp. If you have home field advantage, you know what to do: The folks who water the ground can fix things for you just the way you'd like it. Plan B, if the ground is frequently soggy, is to add heating pipes under the top soil. That way, in addition to evaporating the water via the power of the sun, you can heat the soil from below to get rid of the moisture far more rapidly. You can also turn on the pipes to prevent the pitch from freezing. Mind you, they're not cheap!

Feeling drained

When it does rain, it can rain too much. The surface becomes sodden. Luckily, though, soil is not rock solid. Think of a coffee maker. The granules of coffee are packed together tightly. When you drip water on them, though, the water somehow manages to seep through from top to bottom, carrying with it some of the great-tasting caffeine product that is 'the life blood that fuels the dreams of champions.' In like manner, soil also contains tiny little pathways through which the surface water can drain down to the water table. Unlike coffee, though, the water can take with it the nutrients of the soil (which is why you need fertiliser). It can also take chemicals that have been put onto the grass and bears them away into the drinking-water supply, which is rather a problem.

While Mister Darcy is well-known to anyone who has read Jane Austen's *Pride and Prejudice*, Monsieur Darcy, a nineteenth-century French engineer, is slightly less famous. Henri Darcy worried greatly about drainage, especially when planning a new water treatment for the city of Lyons, home to the rugby club Lyon Olympique Universitaire, a.k.a., the Wolves. Darcy supposed that the tiny little channels in the soil were tubes of a certain size. How would the water flow through them? Darcy came up with the law that bears his name. First he could use some off-the-shelf physics from another French physicist, Jean Louis Marie Poiseuille. Poiseuille had looked at how a viscous liquid would flow through a cylinder. He showed that if a pipe of length L is at an angle, so that its top end is a vertical height H above its bottom end, then the amount of liquid passing through the pipe per unit time, Q, is:

$$Q = \frac{\pi D^4 \rho g}{128\mu} \frac{H}{L}$$

where μ is the viscosity of the liquid, which you can usually look up in a physics, chemistry, or engineering handbook. *D* is the pipe diameter. Darcy noticed that when he packed a pipe with sand, the flow had a discharge per unit time of:

$$Q = KA\frac{H}{L}$$

with *A* the cross sectional area and *K* is a constant that depends on the stuff that's packing the pipe. In a real pitch, suppose there are tiny cracks, pores, or drainage channels that are typically of diameter *d*. Darcy proposes that the hydraulic constant, *K*, is:

$$K = Nd^2 \left(\frac{\rho g}{\mu} \right)$$

Here, N is a number that reflects the effect of shape on the drainage channels. (See Figure 5.3)

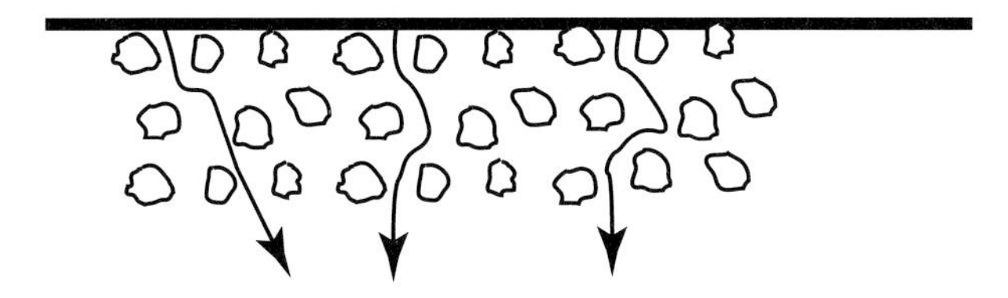

Figure 5.3: Soil is porous, so water can drain through tiny channels, much like a coffee percolator.

If the pitch is dripping wet, there are a couple of things that Darcy's law suggests we ought to do. We have no control over gravity, nor on the density and viscosity of water. We can, though, do something about the tube size. First, you can send troops out onto the pitch with, dare I say, pitchforks. Stamp them down, work the fork around a little bit and then take the fork out. This creates a number of far larger channels, whose radius d is that of the tine, rather than the grain size of the soil.

If this doesn't seem obvious, there's something even more blatant. Geologists know that the hydraulic conductivity depends on the *intrinsic* permeability of the soil, which means that the rate at which water drains through a material depends on the size of the granules of that material. With sand, you can hold the granules in the palm of your hand. If you have clay, you'll be lucky to see the pores under a microscope. So, water should drain more swiftly from sandy soil than from clay soil. If you want to play rugby close to London, which has a substrate of (you guessed it) London clay, then to avoid waterlogging, keep a good supply of fine sand on hand. When the water begins to pool, pour sand on top. The water will be sucked up through the sand pores, as Darcy's law predicts, leaving the claylike soil far less moist than before.

No matter the problems with your local pitch, things probably aren't as bad as when Western Samoa played Fiji back in 1924 – there was a tree on the halfway line! It was probably a great second row – large girth, tall, but not very mobile. For the record, Fiji ran out winners 6-0.

THE ROAR OF THE CROWD

Rugby is a game of strong emotions. The triumphal singing of 'Land of My Fathers' when Wales are in front, 'Swing Low, Sweet Chariot' when England are in the lead, or the chants of 'Allez le France' at the Parc des Princes are all part of what makes the Six Nations Championship so enjoyable. Such songs are ritual chants of defiance whose purpose is to intimidate the visiting team. The best incarnation of this is the All Black's haka, a traditional Maori chant. The first time they used it in Britain was in 1905, Einstein's year of miracles, before a game against Wales. A New Zealand newspaper reported:

> 'The war cry went well and the crowd listened and watched in pleased silence, and thundered their approval at its close. Then the Welsh team started their national anthem. Forty thousand Welsh voices caught up the noble strain, and from every comer of the ground rose the deep, swelling, heart-stirring chorus *Mae hen wlad fy nhadau*'.

The audience in South Wales did better than the players in New South Wales. When the Australians saw the haka for the first time, they complained it was unfair that they had to begin play while still frightened to death. Famously, in 1989, the Irish replied to the haka at Lansdowne Road when their captain Willie Anderson led an arrowhead formation to stare down the New Zealanders, much to the delight of the home crowd. Inspired though they were, the Irish lost 28-6.

New Zealand is no longer alone. Tonga begins its matches with the Kailao, a rival to the haka, which makes for some thrilling, and loud, opening rituals when the All Blacks face the Sea Eagles. How war cries, songs, and random shouts can be heard in a stadium is a branch of physics and engineering called acoustics. Your vocal chords, lodged in the Adam's apple area, move around whenever you try to speak. The air that we breathe has a certain pressure and a certain density. When your vocal chords vibrate, they create a tiny variation in air pressure. This pressure change exerts a force on the air molecules, and causes slight changes in the air's density, too. As a result, the sound goes out of your mouth into the air outside. The sounds you produce have the form of a wave. Rather like a wave of the top of a pond or the sea, these waves have an amplitude, wavelength, frequency and speed.

The speed of sound depends on the air pressure, density, and temperature[4]. It's around 344 m/s at 20 Centigrade at sea level. As the weather gets warmer, the speed increases, by about 0.6 m/s for every extra Centigrade. Humidity also affects sound speed, but it's much less noticeable. At 20 Centigrade, sea level, and a relative humidity of 50%, the speed of sound is 343.99 m/s. At 98% humidity, where you'll be dripping wet almost instantly, it's 344.59 m/s. The finite speed of sound creates a time lag. A vigorous call for the mark on your own goal line will be heard three tenths of a second later by a fullback on the other goal line (though what he would be doing in that position is a mystery).

If the scrum half, pack leader, coach, referee, or diehard fan have something to say or scream at you, how loud do they need to say it? How loud is loud enough? The answers depend on the intensity with which they speak and the attenuation of the sound wave as it travels through the air.

When you speak, your vocal chords impart a certain amount of energy to the air in your mouth. As the sound travels outwards, this energy is spread out over a wider and wider area. When the sound wave has travelled a distance R, for example, the energy you provided is spread out over the surface of a sphere of radius R. In other words, if you provided an energy E to the sound waves, the energy per unit area after the wave has travelled the distance R is $E/4\pi R^2$. This quantity, the energy per unit area, is the intensity, I. So, the intensity $I(R)$ at a distance R from the speaker is, roughly speaking:

$$I(R) = \frac{I_0}{4\pi R^2}$$

Here I_0 is the intensity at some reference distance, say a metre from the shouter or, in the case of the ref, the whistle blower.

If you double the distance, then you reduce the intensity by a factor of four. That's bad news for a winger who is trying to hear what the scrum half is saying. In practice, things are worse. A sound wave is a combination of pressure and density waves. As it travels, part of the sound wave's energy has to go into making stationary air move, and to do that it has to overcome the viscosity of the air. The sound wave also compresses air and, just as the pump gets hot when you blow up a rugby ball, so some of the sound wave's energy goes into heating up the air through which it passes. So, in practice,

$$I(R) = \frac{I_0}{4\pi R^2} \exp(-R / R_0)$$

The exponential function kills off the intensity quickly and usually it's an exponential that's the important killing factor. Not so for sound. The distance over which the exponential decays is about R_0, but for air this is a distance of about a kilometre. Experiments in acoustics show that the range depends on the frequency, so that $R_0 \sim 1/f^2$. In other words, if you double the frequency, you reduce R_0 by a factor of 4, decreasing the range the sound carries significantly[5]. That's why thunder, far off in the distance, is a low rumble – the attenuation prevents the higher frequencies from reaching your ear.

If the scrum half yells, it won't ruin the fly half's hearing. Our ears – as long as Johan le Roux hasn't half-removed them - work on a logarithmic scale. That means that if we hear two sounds, one a 100 times more intense than the other, we hear it as being twice (2 bels) as loud. The best sound level we can hope to hear is 0 bel. Ordinary speech lurks at the 40 to 60 decibels mark, while a jet engine 100 metres away kicks out about 140 db, slightly above the pain threshhold.

The ear, a surprisingly wonderful instrument, can detect frequencies as low as 20 Hertz and as high as 20 kHz. Age, and listening to loud music, will affect these limits. As an example of attenuation, in air at 0 Centigrade, a 125 Hz shout will drop off by about 0.1 dB over a distance of 100m, while a 2000 Hz signal drops off by 2.3 dB over the length of the pitch. Mind you, it depends on humidity – these numbers work at 10% humidity but are cut in half at 90% humidity. The wetter the air, the farther the sound can travel without suffering damage.

Back in 1883, Birmingham instrument maker James Hudson developed a whistle, the Acme Thunderer, which could be used by the British bobby. Today, Acme Thunderer Whistles are still in use and still popular with referees. They put out a sound level of about 119 db. The most powerful whistles in the world, also produced by Acme, are the ones built into aircraft lifejackets in case of a water landing. These usually feature two pitches. The first is high, with relatively short range. It is shrill, and so stands out distinctly from the mere hubbub of background noise. The second pitch is lower, and this one has a far longer range, enabling you to be heard across the stormy waters. The standard issue Acme Thunderer for rugby referees likewise has the shrill tone combined with a pitch that'll easily carry across the hallowed turf of HQ. The high pitch means you can't say you didn't notice it; the low pitch means you can't say you were too far away to hear it. 'Sorry ref, must have been the attenuation' won't work as an excuse.

As play draws to a close, fans of the winning team often start whistling to signal the end of the game. Yet the players never stop. The reason is profound. Any musical note that doesn't last infinitely long – and that's all of them – cannot be pure. In quantum mechanics, this is known as Heisenberg's Uncertainty Principle. To engineers it's a consequence of the bandwidth theorem. An impure note has a certain blend of frequencies in it. Just as one blend of red, white, and blue give lavender, but another one imperial purple, so one blend of a note hits our ears and we say 'guitar' and another, playing the same note, is unmistakably 'piano.' Just like fingerprints are unique, so is the blend of frequencies put out by the Acme Thunderer and the human larynx. So, whistle as loudly as you want, for you might psychologically affect the referee and time keeper; but you'll never fool the players.

The logarithmic scale of hearing helps anyone – player or spectator – at the Millennium Stadium. Singing is a tad louder, usually, than speaking. Suppose Welsh fans are in fine form, and *Land of My Fathers* is belted out by every Welshman at about 75 decibels. There are some 100,000 diehard daffodil-wearing leek-carrying patriots singing. If sound intensity was added up the usual way, rather than logarithmically, eardrums would burst. A sound wave of 7.5 Million decibels would be destructive. Luckily, the entire crowd has a loudness that is the logarithm of 100,000, which is 5. So, the entire crowd sounds about 5 bels louder than a single human voice. Mind you, if one person sings at 75 decibels, the entire crowd amasses 125 decibels, close to jet engine, pain threshold levels. No wonder it's so intimidating to play there. There is one advantage, though: humans absorb sound. The average member of the Cardiff crowd is equivalent to about 4.7 square metres of perfect sound absorbing material. So, there's not much in the way of echo, allowing for the wonderful harmonies of 'Bread of Heaven' from all sides of the stadium.

Hearing is not everything. Referees have been instructed on how to officiate in games when deaf or hard-of-hearing players are participating. Several nations now have rugby teams for the deaf; to qualify for Scotland's, your hearing loss in both ears must amount to about 25 dB. The first Deaf Rugby World Championship was held in Auckland, New Zealand where in the final the Welsh team beat their hosts 28-14. It is a shame, if not an outrage, that the second world championship had to be cancelled for financial reasons.

WILL YOU WIN? REFEREES, QUANTUM MECHANICS, AND THE THEORY OF GAMES

Suppose you wear the physics-specified clothing for a game, the pitch is prepared, and the crowd sings so loudly that it drowns out your opponent's scrum half. When they have the ball, they simply cannot communicate. Can you win? No matter how badly your team usually plays, there's always a chance. You are, though, in the hands of referees, and how they police the game has a profound effect on the final score.

One of the great mathematicians of the twentieth century was John von Neumann, a Hungarian émigré to the United States. He wrote, with Oscar Morgenstern, a definitive book on the mathematics of game theory and another one, all by himself, on the mathematical foundations of quantum mechanics. The first subject, though von Neumann may not have recognised it, explains unsportsmanlike conduct on the pitch, the second, the difficulties of refereeing.

Games, to a mathematician, are a lot different from what most people think of as a game. The simplest game to analyse bears a passing resemblance to the classic game of childhood, 'rock, paper, scissors.' On the count of three, you pick one of those three; so, too, does your opponent. Rocks blunt scissors but are wrapped by paper, while scissors cut paper. Violent children (the future props) can use rock, dynamite, and scissors, for dynamite will blow up rocks but scissors will cut the fuse. No matter how you label the options, the wins, losses, and draws are clear.

For the simplest mathematical game, we have not three options but only two. Label these 'obey' or 'disobey' as in, 'obey or disobey the rules of the game'. Play this game in the scrum, for sake of argument. As they pack down, each eight chooses one of these two options. There are four possible outcomes: both teams obey; both teams disobey; we obey but they disobey; or we disobey while they obey. While I would never encourage teams to ignore the code of conduct set forth by the Rugby Football Union and I have never (hem, hem) transgressed the laws myself, some teams do. Why?

The secret is wrapped up in a reward system. This is most easily pictured in a simple mathematical structure properly known as a pay-off matrix:

$$\begin{array}{cc} & \text{Them} \\ & \begin{array}{cc}\text{obey} & \text{disobey}\end{array} \\ \text{Us} \begin{array}{c}\text{obey} \\ \text{disobey}\end{array} & \begin{pmatrix} a & b \\ c & d \end{pmatrix} \end{array}$$

This tells us, in mathematical form, that if both teams obey the rules, our team earns a reward a. If they disobey the rules while we obey, we earn b. If we disobey the rules but they obey, our pay is c, and if both teams are naughty, we earn d. If any of these values is less than zero, say -5, it means that they get a reward of 5. To be cold blooded about things, if our disobey row adds up to more than our obey row, $c+d > a+b$, we will have a greater reward for breaking the rules of the game than for obeying them. If $a+c>b+d$, they earn more for obeying the rules than disobeying them.

In an ideal world, we all obey the rules all the time. The problem, though, is if the other team is better than you. If their hooker perpetually heels the ball against the head, doesn't your hooker sometimes want to risk raising his foot before the ball comes in?

Game theory explores the outcome of strategies, to find out what happens if we obey the rules a fraction x of the time and disobey $(1-x)$ of the time. The other team, we think, obeys a fraction y of the time and disobeys $(1-y)$ of the time. Fairly simple arithmetic[6] dictates what strategy is best for both teams. We should select:

$$x = \left(\frac{d - c}{a + d - (b + c)} \right)$$

while they should follow the rules:

$$y = \left(\frac{d - b}{a + d - (b + c)} \right)$$

There is one last mathematical sting in the tail, which is the so-called value of the game. This is the reward that *our* team can expect to get if both teams play their best strategies. For our case, the value v is:

$$v = \frac{ad - bc}{a + d - (b + c)}$$

What does this have to do with scrums? Almost everything. All we have to do is to model rewards and punishments. Suppose when we both play by the rules, we have a good chance of beating them in the scrum, and thus set $a=2$. If we play fairly, but they engage – heaven forbid – in foul play, perhaps they would win the scrum more often than we would. Our reward is -1 (or, put another way, their reward is 1). In contrast, if we cheat while they play fairly, suppose our reward is -1.5. (For some reason, it's much more obvious when our hooker lifts his foot than his opposite number so we are penalised more frequently). Last, suppose that if both teams ignore the rules, the ref lets it go, but now they have a better chance of winning the scrum. So our reward for a fair fight is +1. The pay-off matrix for this game is:

$$\begin{pmatrix} +2 & -1 \\ -1.5 & +1 \end{pmatrix}$$

Plugging in the numbers, mathematics tells us to rough it up in the front row an astonishing 55% of the time; they should do so 63% of the time. The value of this game is 0.09, which is greater than zero. Our team, therefore, has a positive reward from the scrum experience. Given the famous choice of 'kick again, line, or scrum' when the ball sails out after a restart, we'll have the scrum, please, ref.

Sad to say, this whole line of mathematical reasoning sums up what those of us who toil in the front row have long since known: if you easily beat your opponents, they'll soon start the 'trash talking' and other behaviors too appalling to be described in a family book.

To see what effect a ref can have, imagine the following. The two teams are evenly matched when both play fairly or when they both engage in skullduggery. In that case, $a=d=0$. The ref blows the whistle and savagely penalises any unilateral bad behavior. Our reward for playing fairly when they cheat is +2, say, and the ref – a bastion of equality – rewards us by -2 if we cheat while they obey the rules. The payoff matrix is then:

$$\begin{pmatrix} 0 & 2 \\ -2 & 0 \end{pmatrix}$$

The value of this game is 0, so neither team benefits, or loses, from the scrum. But both teams benefit from the same strategy: The only one that makes sense, thanks to the ref, is to play by the rules. The man in the middle can stop tempers from fraying and save people from injuries by dispensing justice in the form of free kicks, penalties, and cards of the yellow and red persuasions. Anything else is irresponsible.

In the subatomic world, elementary particles can behave bizarrely. That, too, was summed up by John von Neumann. But bizarre behavior is not confined to the quantum world - some referees behave pretty oddly, too. In the realm of the quantum, it is impossible to know precisely where a particle is and, at the same time, know how much momentum it has. It's not an issue of not having decent enough measuring sticks, it is simply impossible. Formally, this is known as Heisenberg's Uncertainty Principle. It covers more than just position and momentum; it says you can't know how much energy something has and, simultaneously, know when it has it, for energy and time are another such pair.

Philosophers of science like to think of this as something deeply profound. In a way, it isn't. The same piece of mathematics lets you know that a musical note cannot be pure unless it lasts for an infinite time (energy and time). Rugby analogies are the ruck, maul, and scrum. Let's face it, the ref has a choice. He can either make you stay on your feet and keep your paws out of the ruck, or he can stare down the backs to make sure they don't encroach offside. The two are not simultaneously knowable.

There's another parallel between the quantum and the referee. Electrons, as they whiz around the nucleus, are only allowed certain energies. In the jargon, they can only be in a certain number of states. The curiosity is that, when you ignore the electron, it can be any mixture of these states that you can think of. If you measure some property of the electron, though, it will be in precisely one of these states. The experts say that the quantum mechanical measurement process has 'collapsed the wave function' of the electron. To return to our man with the whistle, suppose he's looking at the backs. In the ruck, who knows what's happening? There are two possible states, good and evil. The referee, back to the ruck, knows there's probably a blend of nice and nasty going on behind him – it's a mixture of those two states. He spins around, makes a measurement, and either blows the whistle or not. By looking at the ruck, the ref has forced it to collapse, if you'll pardon the expression, either into a law-abiding state or one of lawlessness. Usually, it's the latter.

Quantum physics is described by probabilities. The chance that an electron is here, or the chance that two protons will manufacture a Higgs boson. In rugby, chance plays a role too. What are the chances, then, of victory? Are there any lessons that probability can teach us?

Suppose the other team is 5 times better than you are. Technically speaking, if you played six games, they would win five and you would win only one. But you're not playing six games, just one. The odds may be in their favour, but you *always* stand a chance. Never give up until the final whistle has blown, and never let a previous season's drubbing lead you to think you'll *never* beat that team.

In years gone by, there was a Five Nations championship, where England, Ireland, Scotland, Wales, and France battled for Northern Hemisphere dominance. What would happen if all the teams were evenly matched? Our put another way, let's assert that all teams are equally good. What would happen? To win the Grand Slam, you had to win four games. The chance that you win the first game is one half. The chance you win the second game is also one half, if you and your opponents are equally matched. So, you have a 1 in 4 chance of winning the first two games. By extension, France had a 1 in 16 chance of winning the Grand Slam in any given year. France was just one of five nations, though. The chance that someone wins a Grand

Slam is going to be 5/16, which is about 31%. A swift look at the Five Nations championships between 1947 and 1999, the year before the Azzurri were added to make it the Six Nations, shows something intriguing. There were Grand Slams in 18 out of the 51 years in which the tournament was completed. In other words, there was a Grand Slam winner in 35% of the time. Our model, all the nations are evenly matched, looks good.

To win the Triple Crown, you have to win three games, which your country has a 1 in 8 chance of achieving. As there are four nations competing, the chance that one of them wins the Triple Crown is going to be 4/8, or one half. The first year that teams from the British Isles competed against each other was in 1884. On January 5th of that year, England opened its campaign with a 5-3 victory at home over Wales, beating Ireland 3-0 on February 4th, and concluding with a 3-1 victory at home against Scotland on March 1st to secure what we would now call the first ever Triple Crown. In the past 51 years, there have been 22 Triple Crowns earned, which is about 43% of the time. This simple model suggests that the Five Nations teams were indeed fairly evenly balanced – the chance that you can wear your country's jersey with pride after the Six Nations championships does not depend on the country you support! Mind you, the opposite of victory is defeat. If your chance of winning any individual game is 50-50, that's also your chance of losing every individual game. There was a time when the Welsh sun was in eclipse and they lost 8 out of 9 games, their sole victory being over the tiny island nation of Western Samoa. As Gareth Davies wryly remarked, 'good job we didn't play the whole of Samoa.' Fair play, though — the Welsh are non-starters at losing streaks. Among the Six Nations, the French hold the record, racking up eighteen in a row from 1911 through to 1920 narrowly edging out Scotland which, with 17 losses in the period 1951-55, merits a dishonourable mention.

At the end of the day, as you head for the train, the bus, or walk the long journey home, there's plenty to think about. As usual, there are the great moments to savour, when you, your teammates, or the team you cheer for have done wonderful things. There is also the long silent walk home reflecting upon the anguish of defeat or the embarrassing play. But now, there's a new angle. The 80 minutes of rugby was also a full 80 minutes of physics. There were moments, no doubt, where Newton and Co. were dictating the play. Before the game fades from memory, what examples of physics in action could you see? And how can you capitalise on them in next week's game? If Einstein is your coach, how can you lose?

CHAPTER FIVE - ENDNOTES

1 As $K = 4\sqrt{\left(\frac{H}{V}\right)} + \frac{2}{H}$ and V is fixed, we get:

$$\frac{\partial K}{\partial H} = \frac{4}{\sqrt{V}}\frac{\partial}{\partial H}\left(\sqrt{H}\right) + \frac{\partial}{\partial H}\left(\frac{2}{H}\right) = \frac{2}{\sqrt{VH}} - \frac{2}{H^2}$$

This is zero when $\sqrt{VH} = H^2$ or $H^{3/2} = \sqrt{V}$

2 This result is not a fluke. Suppose we model players as cylinders instead of cuboids. The surface area of a cylinder of height H and radius R is $A = 2\pi R^2 + 2\pi RH$. The volume is $V = \pi R^2 H$. Thus

$$K = \frac{2\pi R^2 + 2\pi RH}{\pi R^2 H} = \frac{2}{R} + \frac{2}{H}$$

For constant volume,

$$K = \frac{2}{H} + 2\left(\frac{\pi H}{V}\right)^{1/2}$$

So, for two players of equal mass, the one with $R = H/\sqrt{2}$ will have the lowest K value.

3 Keen readers of the medical literature will no doubt recall the article by Messrs D. and E.F. Dubois, *A formula to estimate the approximate surface area if height and weight be known*, which was published in the Archives of Internal Medicine 17 863-871 (1918). They reported that the surface area S depended on height H and mass M as: $S = 0.007184M^{0.425}H^{0.725}$. ($S$ is in metres squared, H is in cm, M is in kilograms). In their model, the K value, S/V, would be proportional to $S/M \propto (H^{0.725}/M^{0.575})$. Using this formula, K(Adam)/K(Daisuke)~ 0.798. Our approximate formula in the chapter is 0.82.

Given the Dubois formula, the model of humans as cylinders works better than humans as cuboids. We know Adam's mass and height and how dense he is (physically speaking, I add hurriedly), and so can work out what the radius is for a cylinder of this height and mass. Then we can work out the surface area. We can do the same for the Adam as cube model. The Dubois formula gives S=2.45 ms^2; the cube model gives 2.06, and the cylinder 2.63. For Ohata, the numbers are 1.81, 1.45, and 1.93 respectively.

4 The speed of sound in an ideal gas is $c = \sqrt{(\gamma\, p / \rho)} = c_0 \sqrt{T / 273}$. Here γ (gamma) is the ratio of specific heats, p the pressure, ρ (rho) the density, and T the air temperature (in Kelvin). c_0 is the speed of sound at 273K. On a hot day, when T = 300K, the speed of sound differs from its 273K value by a mere 5%.

5 For completeness, the attenuation goes as:

$$\frac{1}{R_0} = \frac{2\pi^2 f^2}{\rho_0 c^3}\left[\frac{4\eta}{3} + \left(\frac{\gamma - 1}{\gamma}\right)\frac{M\kappa}{C_V}\right]$$

Here η is the viscosity, M the molecular weight, κ the thermal conductivity, and C_V the molar specific heat at constant volume of air. The major determining factor is the frequency, f. Higher frequencies have shorter ranges than lower frequencies.

6 If they always choose to obey while we obey x times and disobey 1-x times, our reward is $ax + c(1-x)$. On the other hand, if they always disobey, our reward is $bx + d(1-x)$. So, if they are honest y times and dishonest 1-y times, our entire reward package is $y[ax + c(1-x)] + (1-y)[bx + d(1-x)]$. Our best strategy is one that doesn't depend on what they choose to do. This means that the reward for our best strategy can't depend on y. That happens if $[ax + c(1-x)] - [bx + d(1-x)] = 0$ or, equivalently:

$$x = \frac{d - c}{a + d - (b + c)}$$

Suggested further reading

Biographies

Many of the great players mentioned in this book have written, co-written, or have had ghost-written biographies published. Here are a few:

Allan Bateman and Paul Rees *Allan Bateman: There and Back Again* (London: Mainstream Publishing, 2001)

A fantastic player and fearsome opponent, Bateman represented Wales in both Rugby League and Rugby Union

Phil Bennett and Graham Thomas *Phil Bennett: The Autobiography* (London: HarperCollinsWillow, 2003)

The story of arguably the greatest fly half ever.

Alice Calaprice and Trevor Lipscombe *Albert Einstein: A Biography* (Westport, CT: Greenwood, 2005)

An accessible guide for non-physicists into the life and science of the world's most famous physicist

Ray Gravell and Alun Wyn Bevan *Grav in his Own Words* (Llandysul: Gwasg Gomer, 2008)

Welsh-language soap opera star, rugby playing phenomenon, and TV commentator are only some aspects of the intensely patriotic Gravell, surely one of the most-loved players ever to wear the red of Wales.

John W. Keddie *Running the Race: Eric Liddell* (Darlington: Evangelical Press, 2008).

A fascinating biography of the Flying Scotsman, made famous by 'Chariots of Fire', who won Olympic gold for the United Kingdom and played on the wing for Scotland. He spent most of his life as a missionary to China.

Jonah Lomu *Jonah Lomu: The Autobiography* (London: Headline Books, 2004).

No awards for the originality of the book title and subtitle, but it is the story of one of the most formidable players to play for New Zealand.

Jean-Pierre Rives and Roger Blachon *Vestiaires* (Paris: Editions Anne Carriere, 2007)

Reflections and opinions of one of France's most spellbinding players.

JPR Williams *JPR: Given the Breaks – My Life in Rugby* (London: Hodder and Stoughton, 2006)

How legendary BBC commentator Bill McLaren listed Andy Irvine at fullback instead of JPR in his World Rugby XV is a complete mystery to all those whose name is not Bill McLaren. I bet even Andy Irvine was shocked.

Shane Williams and Delme Parfitt *Shane: My Story* (London: Mainstream Publishing, 2008)

No matter whether your watching the game on TV or live, you can still here the crowd shout 'give it to Shane.' One of the most exciting players ever.

Richard S. Westfall *Isaac Newton (Very Interesting People)* (Oxford: Oxford University Press, 2007).

A bite-sized look at Newton by an author who wrote the definitive biography (*Never at Rest*, published by Cambridge University Press)

THE SPORT OF RUGBY

Michael Green and John Jensen *The Art of Coarse Rugby* (London: Robson, 1998).

Laugh-out-loud funny. This sums up how the game is truly played, rather than the way the England selectors think it is.

John Griffiths *Rugby's Strangest Matches: Extraordinary but True Stories from over a Century of Rugby* (London: Robson Books, 2000)

Fun reading during the off season.

Huw Richards *A Game for Hooligans: The History of Rugby Union* (London: Mainstream Publishing, 2006)

No home should be without a history of rugby. This is an enjoyable one.

Derek Robinson *Rugby: A Player's Guide to the Laws* (London: HarperCollinsWillow, 2005)

Let's face it, players and referees need all the help they can get. The laws of the game as published by the RFU can only appeal to those who regard legal contracts and rental agreements as bedtime reading.

PHYSICS

George B. Benedek and Felix M.H. Villars *Physics, with Illustrative Examples from Medicine and Biology Volume 1: Mechanics Second edition* (New York: Springer, 2000)

A book that touches on many of the topics included in this book. It covers bone fractures in some detail.

C.B. Daish *The Physics of Ball Sports* (London: English Universities Press, 1972).

A classic book that includes discussion of the prolate spheroid we know and love.

B. H. Flowers and E. Mendoza *Properties of Matter* (John Wiley: Chichester, 1970).

A lovely treatment of solids, liquids, and gases. It covers, at a slightly higher level than this book, random walks in gases.

David Griffing *The Dynamics of Sports: Why that's the Way the Ball Bounces, Fourth Edition* (Oxford, Ohio: Dalog Company, 1999).

An introductory book that looks at many of the topics relevant to rugby, such as projectile motion in ball sports; running; and tackling (albeit for American rather than rugby football).

Göran Grimvall *Brainteaser Physics* (Baltimore: Johns Hopkins University Press, 2007).

A pleasurable book that provides superb insights into diverse physical systems.

Paul Hewitt *Conceptual Physics* (New York: Scott Forsman Addison Wesley, 2008)

A standard university-level books that provides an introductory overview of physics.

Bruce Schumm *Deep Down Things: The Breathtaking Beauty of Particle Physics* (Baltimore: Johns Hopkins, 2004)

Those intrigued by the discussion of Feynman diagrams and their odd connection to passing, interceptions, and tackling, will enjoy this book. It looks at one of the great accomplishments of the twentieth century, the standard model of particle physics, and uses cartoons and descriptions to help a general reader understand what all the fuss is about.

Colin Smith *This Cold House* (Baltimore: Johns Hopkins, 2007).

Aimed at those who want to know how to keep their homes warm and comfortable, the physics relates to much of chapter 5 in terms of heating, cooling, and evaporation.

Clifford Swartz *Back of the Envelope Physics* (Baltimore: Johns Hopkins, 2003)

This should be compulsory reading for physicists. Great training in getting 'good enough' estimates of physical systems.

Jason Zimba *Force and Motion: An Illustrated Guide to Newton's Laws* (Baltimore: Johns Hopkins University Press, 2008)

A look at Newton's laws with an emphasis on qualitatively understanding what they mean. This is good background reading for chapters 1-4.

MATHEMATICS

John A. Adam *Mathematics in Nature: Moedling Patters in the Natural World* (Princeton: Princeton University Press, 2003)

Written at a higher mathematical level than this book, Adam discusses several of the topics in greater detail and in an engaging prose style. He looks at surface area to volume ratios, for example, as well as Strouhal numbers and vortex sheeding.

Robert B. Banks *Towing Icebergs, Falling Dominoes, and Other Adventures in Applied Mathematics* (Princeton: Princeton University Press, 1998)

Bob Banks was a master of fascinating mathematics problems and for his services to engineering received the Order of the White Elephant from the people of Thailand. His book includes extensive use of differential equations. The Hill-Keller models of running (in chapter two of this book) is covered in detail, as is the fall of a tall rod, which is described (at far lower mathematical level) in the section of this book where Jonny Wilkinson is tackled.

Keith Gregson *Understanding Mathematics* (Nottingham: Nottingham University Press, 2007).

This short book covers, in a clear accessible way, all of the mathematics used in the main text of this book.

Bart Holland *What are the Chances: Voodoo Deaths, Office Gossip, and Other Adventures in Probability* (Baltimore: Johns Hopkins University Press, 2002)

This is an easy-to-read book that covers – at about the same level as this book – some offbeat topics where chance and probability play an important role.

Marshall Jeavons *Murder at the Margin (A Henry Spearman mystery)* (Princeton: Princeton University Press, 1993).

Not the most eloquently written murder mystery, but the protagonist is an economics professor who, on vacation, solves a murder. To do so, he employs game theory, which is mentioned in this book in chapter 5.

Paul J. Nahin *Chases and Escapes: The Mathematics of Pursuit and Evasion* (Princeton: Princeton University Press, 2008)

An enjoyable read that covers, at a higher mathematical level than this book, how to chase effectively.

M.S. Makower and E. Williamson *Operational Research* (London: Hodder and Stoughton, 1975).

I bought this book in a second-hand shop many years ago. It is well-written and covers some basic elements of probability theory. In this book, it covers much of the material in game theory (co-operating or not in the scrums) covered in chapter 5.

Lucy Tucker *Simplistic Statistics* (Lincoln, Chalcombe Publications, 2003).

A concise guide to some elements of statistics, such as those used in the discussion of tackling in this book.

INJURIES

N. Nichole Barry, Michael F. Dillingham, and James L. McGuire *Nonsurgical Sports Medicine* (Baltimore: Johns Hopkins University Press, 2002).

How to deal with niggling knee injuries, and so forth.

Sara Palmer et al. *Spinal Cord Injury, Second Edition* (Baltimore: Johns Hopkins University Press, 2008)

If you are a coach or referee, it would be morally irresponsible not to read this book. It provides the latest advice from the medical staff of America's best hospital on how to cope with injuries that occur when, for example, a scrum or maul collapses.

GLOSSARY

Acceleration ($\vec{a}$). This is the rate of change of velocity with time, so that $\vec{a} = d\vec{v}/dt$. It is a vector and has units of m/s^2. The acceleration due to gravity here on the Earth's surface, for example, is about 9.8 m/s^2.

Angular momentum ($\vec{L}$). The angular momentum of a particle about a point is the mass of the particle multiplied by its distance from the point and by its speed. It's direction is perpendicular to both. In vector notation, $\vec{L} = m\vec{r} \times \vec{v}$.

Angular velocity ($\vec{\omega}$. The Greek lower-case letter omega). If a particle moves with velocity $\vec{v}$ and is at a position $\vec{r}$ relative to some origin, then the angular velocity $\vec{\omega}$ of the particle is $\vec{v} = \vec{\omega} \times \vec{r}$. The SI units of angular velocity are radians per second.

Component A vector, which has a size and an associated direction, can be split into two separate components pointing in two different directions. For example, a force acting in the direction 'North East' can be thought of as the sum of two forces, one in a northerly direction, one in an easterly direction. That same force can also be thought of as the sum of a force headed south south west and one pointing east north east.

Conduction Thermal conduction is a process where heat energy is transferred through a body by direct contact, rather than by motion within the body itself. The flow is from hot regions to cold. How well a given material conducts heat is measured by its thermal conductivity. At 25 Centigrade, the thermal conductivity for aluminium is about 250 W/mK, while that of rugby-jersey cotton is a mere 0.3 W/mK.

Convection A process in which heat is transferred through a liquid or gas by motion within the fluid itself.

Density The mass per unit volume of a substance. The units are kg/m^3. Water has a density of 1000 kg/m^3. Air, on the other hand, has a density of 1.3 kg/m^3. Traditionally, density is represented by the Greek letter rho, ρ.

Diffusion The spreading of molecules of one type in a medium of another type (if the two types are the same, this is self-diffusion). These ease with which a substance can diffuse in another is summarised by the diffusion coefficient, D, whose units are m^2/s.

Drag force The resistive force exerted by a fluid on a moving object immersed in it.

Energy is the capacity of a body to do work. A scalar quantity, it is measured in Joules (J), where 1J = 1 N.m or 1 kg m^2/s^2. Potential energy is that due to a body's position, and is usually connected to the gravitational potential energy.

Force ($\vec{F}$) A force is a push or a pull, something that speeds up, slows down, or changes the direction in which an object is moving. From Newton's Second Law, it is the rate of change of momentum, $\vec{F} = d\vec{p}/dt$, and is measured in Newtons (N), which are kg m/s^2.

Friction is a force that resists the motion of an object that moves over a surface. If the force normal to the surface is N, then the friction force is fN, where f is the coefficient of static friction. The coefficient depends on the two surfaces involved, but not on the surface area of the object. Values of f are between 0 (ice) and 1 (thick mud).

Heat capacity This is the amount of energy required to raise a given amount of a substance by one degree Kelvin. The heat capacity depends on the thermodynamic process involved, for the specific heat capacity at constant pressure differs from the heat capacity at constant volume. For air, the specific heat capacity at 0 Centigrade (at sea level) is about 1 kJ/kg K. Ice, at -10 Centigrade, is about double that.

Latent heat The latent heat of *fusion* is the amount of energy required to transform a given amount of a substance from its solid form to liquid. The latent heat of *vaporisation* is the amount of energy needed to transform a given amount of a substance from the liquid to gaseous form. For water, the latent heat of fusion is 334,000 J/kg. or 334 kJ/kg.

Mass The amount of 'stuff' of which something consists. It is measured in kilograms and reflects the number and type of molecules of which an object is made.

Moment of Inertia (I) measures how easy it is for a physical body to be rotated by a given torque. If a particle of mass m at the end of a string of length L is to be rotated about the end of the string, its moment of inertia is mL^2. If an object is made up of N masses, the mass of the i'th particle being m(i) and its distance from the centre of rotation being r(i), then the moment of inertia is $I = m(1)r^2(1) + m(2)r^2(2) + \cdots m(N)r^2(N)$. The moment of inertia, therefore, depends not only on the mass of the object but on the way in which that mass is distributed.

Momentum ($\vec{p}$). Linear momentum $\vec{p}$ is the product of an object's mass m and velocity $\vec{v}$; $\vec{p} = m\vec{v}$.

Power (P) is the rate at which work is done and is measured in Watts. If a force $\vec{F}$ displaces an object by an amount $\vec{d}$, then the work done is $\vec{F} \cdot \vec{d}$. If the force is constant, the power is $P = \vec{F} \cdot \vec{v}$.

Pressure (P) is a force per unit area acting on an object, in a direction perpendicular to the object's area. It is measured in Pascals, and 1 Pascal is a pressure of 1 Newton per square metre. Atmospheric pressure is approximately 101,325 Pascals.

Radiation is energy emitted in the form of electromagnetic radiation. Unlike thermal conduction and convection, radiation can work in a vacuum.

Torque ($\vec{N}$) is what causes an object to rotate. If a force $\vec{F}$ acts at a position $\vec{R}$ relative to a turning point (or fulcrum), it generates a torque $\vec{N} = \vec{R} \times \vec{F}$.

Velocity ($\vec{v}$) is the rate of change with position ($\vec{r}$) with time. Hence $\vec{v} = d\vec{r} / dt$.

Work is done when a force $\vec{F}$ displaces an object by an amount $\vec{R}$. The work then done is $\vec{F} \cdot \vec{R}$. Work is measured in Joules.

Player index

(Club affiliation is given if the player is not an international)

INDEX

About the author

Trevor Lipscombe obtained a degree in theoretical physics from London University before going on to earn his doctorate in physics at Oxford. He has taught at Oxford, the City University of New York, and the Johns Hopkins University. He is a Fellow of the Royal Astronomical Society and has published widely in physics research journals as well as journals devoted to physics teaching. Over the years he has taken part in many activities to help bring physics to a wide audience, including such events as an edible-car contest and a mass experiment to see if toast lands butter-side down. In addition to attracting local politicians and the media, these events brought simple science to hundreds of children.

Trevor played rugby throughout high school, mostly in the pack, and played rugby at Oxford until a fateful game when a scrum collapsed and broke his neck. Fully recovered, he now makes his home in Baltimore, Maryland, with his wife and children who - selectors please note - are eligible to play for England, Ireland, Italy, or Wales.